Edwards Air Force Base

Open House at the

USAF Flight Test Center

1957-1966

A Photo Chronicle of Aircraft Displayed

Robert D. Archer

Schiffer Military History
Atglen, PA

Acknowledgments

The scope of this work was made possible by many persons who made it possible for the author to collect the data, take the photographs, and assemble a comprehensive record of the Armed Forces Day Open Houses run by the Edwards AFB personnel between 1957 and 1966. There were many changes during those ten years, but each year people were always willing to help me get to these Open Houses; these included various members of the then new American Aviation Historical Society. Particularly helpful were Harry Gann, John J. Sloan, William Ripley, and Carl Friend.

The staff of the USAF Flight Test Center Office of Information were always very helpful, including Major Ruth M. Dudley, Captain Donald G. Rhoads, 1st Lieutenant B. Kim McDonald, and SSGT George A. Bates.

In addition to his annual visits, the author also worked at Edwards AFB for Northrop Aircraft on several projects. Although unable to carry a camera or take photographs on those occasions, he was able to have many fascinating conversations with many people at the base, including many very well-known test pilots. He is indebted to Northrop test pilots Lew Nelson and Hank Chouteau, for introducing him to many other pilots of the fraternity. The resulting stories are not always printable, but are reflected in the text of this work. Some of the other test pilots the author had occasion to consult with included Joe Cotton, Fitzhugh Fulton and Al White, all of the XB-70 program, and Milt Thompson of the NASA Dryden test center at Edwards AFB.

Many friends have reviewed the draft of this work and made various inputs adding to the quality of the work, besides encouraging the author to persist with the effort; these include Gerald M. Markgraf, the late Richard L. Sloan, and T.R. Waddington. Bob Biondi, of the Schiffer publishing house, encouraged the author to publish the work.

Gerald Markgraf also loaned the author a very valuable publication covering the history of the numerous USAF fighter programs between 1945 and 1973.

Above all, it is the untold thousands of persons, both civilian and military, who work at Edwards AFB, ensuring that all USAF aircraft are tested thoroughly, that made this book possible.

However, the greatest help and encouragement for the entire effort came from my wife, Patricia, who never complained about having to get up at the crack of dawn, allowing me to get mostly crowd-free photographs. My son, Victor, born in 1961, grew up with the noise, smell and sight of modern jet aircraft during the annual visits to Edwards. He is now a very capable critic, and has helped to maintain the photographs at their original quality. All of the photographs in this work were taken by the author, unless otherwise credited.

To each and every one of you, my thanks.

Book Design by Ian Robertson.

Library of Congress Catalog Number: 98-86278

Printed in China.
ISBN: 0-7643-0689-8

We are interested in hearing from authors with book ideas on related topics.

Published by Schiffer Publishing Ltd.
4880 Lower Valley Road
Atglen, PA 19310
Phone: (610) 593-1777
FAX: (610) 593-2002
E-mail: Schifferbk@aol.com
Please visit our web site catalog at www.schifferbooks.com
or write for a free catalog.
This book may be purchased from the publisher.
Please include $3.95 postage.
Try your bookstore first.

Contents

Introduction

Edwards AFB and the aircraft tested there, form a mecca for all kinds of aviation enthusiasts, and the author was delighted when he heard that the base was open to the public once a year. The base was named in 1950 for the USAF pilot killed in the crash of a Northrop YB-49 jet bomber. However, the original name of the site was Muroc which was the reverse spelling of the Corum family name who were the first settlers in the area. The name change to Muroc was caused by the US Post Office asking the Corum's for a change from their original request, for a post office at their settlement to be named after them; this was to avoid confusion with another existing site. I was lucky enough to be introduced to Mrs. Effie Corum Pelton (as she was named then) at the 1960 Open House. Still very active for her age, she told me that her chicken house was now under the site of the biggest hanger at Edwards!

Located out in the desolate Mojave Desert in Southern California, Muroc has always been a remote, difficult place to get to, either as a visitor, or for one working there. This was most certainly true on the occasion of my first visit to the base in 1957 for the annual Armed Forces Day Open House. In the 1950's and 1960's this was always in mid-May and the weather was beginning to warm up to the eighty and ninety degrees Fahrenheit range. Located some 100 miles northeast of downtown Los Angeles, Edwards could only be reached by two or three lane roads for most of the way, (until the freeway system extended out to the Antelope Valley in the 1970's.) This included driving over the Santa Monica mountains on a hair-raising, very twisting highway, with several sheer drop-offs, during the usual Friday night rush. Sepulveda Boulevard then continued almost due north through the San Fernando Valley, before becoming a two-lane road through the mountains to Palmdale and Lancaster.

Many people chose to spend the night at Lancaster, in one of about five major motels, located next to the highway. The adjacent Southern Pacific railway freight trains seemed to run all night long in those days, and woe betide you if you got a room anywhere near the front of the motel!

Up at the crack of dawn, an early breakfast, then one had to drive the final thirty miles through the desert on two-lane roads to the south gate at Edwards AFB, still many miles from the ramp area. If you didn't get up by 6 a.m., you faced the prospect of trying to photograph the display aircraft with hundreds of people around them. The base didn't open for the Open Day until 10 a.m., but the early birds were usually allowed to park near the main ramp entrance and get out onto the ramp prior to arrival of the crowds. As a result, some of the photographs show the aircraft being towed out to their display spots.

I always found it necessary to go around the entire ramp display twice, first to photograph the aircraft, then to record the designations stenciled on the aircraft data blocks (on the left nose of USAF aircraft, or on the lower aft fuselage on all US Navy aircraft). This took a lot of time, and several times I was accused of "spying" for taking such notes. Part of the hazards of the "Cold War"!

The flying display started normally at about 1 p.m., but the runway at Edwards is so far from the ramp and the public viewing area, that it was hard to get any good photographs. The fly-pasts were also a long way from the crowd, quite unlike those at Farnborough in England, or the Paris Air Shows. However, the aircraft, after they had landed, would assemble at the west end of the runway area, and then taxi past the crowd in close proximity, giving one the opportunity to get good photographs.

Normally, the aircraft displayed statically did not take part in the fly pasts, but occasionally a few did. I have given data on all of the static aircraft displayed, plus those taking part in the fly past whose identity could be positively established by the serial numbers on the vertical tail for USAF aircraft. Unfortunately, most of the Navy aircraft in those days had very small serial numbers and these are not readable at the distance they taxied past. A few aircraft in this time period were only in the fly-past and taxi-by, the most prominent one being the second XB-70A, serial number 62-0207, which was seen only at the 1966 Open House, a few weeks prior to its tragic crash.

I was very fortunate to perform on several test and development programs at Edwards during the 1957-1966 time period. These included the Northrop YT-38A spin recovery tests, and the drag parachute tests on the Northrop N156F prototype. Details of these will be found in the Appendices. In addition, I was able to fly two jet aircraft out of Edwards AFB, the Sud Aviation Caravelle VII and the North American T-39 Sabreliner. I have also included a flight in the T-38A out of Palmdale, as it operates almost as an annex to Edwards. Similarly, I have covered the roll-out and first flight of the North American XB-70A-1, from Palmdale to Edwards.

Many very fascinating aircraft were tested at Edwards during this time and I have tried to give the "flavor" of those days in my captions, to give the reader a sense of what it was like there some forty years ago. Working at Edwards was not always easy; the weather could turn windy, blowing fine grit for miles, and in the winter freezing temperatures made the drive to Lancaster a rather exciting one on the often icy roads. I had to make one such trip with a borrowed VW "bus". Its lack of any heating, and engine weight at the rear, made it a real cold and scary drive.

However, there were compensations. On the N-156F drag chute project, I had to take the prototype chute from Irvin Parachute Company up to Edwards and re-pack it to ensure it would work properly. This required a heavy weight at the aircraft end of the shroud lines to hold them true and straight while folding the chute in the motel room. With the enthusiastic help of the Irvin Swedish-born representative, we decided that the only readily available solution was to use a large bottle of liquor from the near-by liquor store. A bottle of Scotch whisky proved to be just the right weight for the job. After packing the chute, we decided to sample the contents of the bottle. A very good evening ensued!

Many problems arose during the development of the YT-38A spin recovery system, but we managed to solve them all and it proved to be very rewarding when the system duly recovered the airplane on the first occasion it was required. Most unexpected was that, after returning to Hawthorne in the company Beech 18, I received a big kiss and hug from the Engineering VP's secretary when I got back to the plant. She explained that she was engaged to the test pilot and had been very relieved to hear that the aircraft had recovered safely.

The major expansion of USAF air power in Vietnam after the Gulf of Tonkin incident in August 1964, led to a major expansion of aircraft and systems of all kinds being tested at Edwards, but this did not get seen publicly until the 1966 Open House. Of course, the actual work had been going on for at least a year before then, but this explains the large jump in aircraft displayed in 1966, plus the re-introduction of camouflage on USAF combat aircraft, such as the F-100 and F-105.

A few new commercial aircraft were also tested at Edwards during the period of the book, both the Douglas DC-8 and the Convair 880 being displayed publicly, but Boeing preferred not to have its 707 seen publicly at Edwards.

It has to be mentioned that not all aircraft that were actually tested at Edwards, were seen at the Open Houses. For example, the Chance Vought XF8U-3 Mach 2 plus fighter was not seen at any Open House, as it had lost the competition to the McDonnell F4H-1 and went back to the manufacturer before the next Open House. However, I have tried to give as complete a picture of the aircraft displayed at Edwards as is practicable; as it is, more than 600 color photographs (all taken by the author) appear herein.

1957

This was my first visit to Edwards AFB, so I had no idea of what to expect. I was delighted to find that two of the original X-series aircraft were displayed, together with one of the Douglas D-558-2s. Other surprises included a Chance Vought Regulus II missile, a Boeing B-17 engine test bed, a Boeing B-29 carrier aircraft, and a North American B-45 bomber. The remainder of the aircraft displayed, both statically and in flight, were most of the major recent types serving with the USAF and the U.S. Navy, or about to enter service. This included the North American F-100, McDonnell F-101, Convair F-102 and TF-102, Lockheed F-104, and Chance Vought F8U. Earlier service types still undergoing testing of some kind, or in use for training and chase purposes, included the Lockheed F-80, Republic F-84, North American F-86, Northrop F-89, Douglas F4D, McDonnell F3H, plus a variety of transport aircraft.

It became evident that most of the major aircraft in the Air force and Navy passed through Edwards AFB Flight Test Center, and that future visits on a regular basis would prove to be most rewarding for any aviation historian. A list of the aircraft displayed this year follows.

Manufacturer	Type	Popular Name	Aircraft Serial
Douglas	A3D-2	Skywarrior	138905
Douglas	A4D-2	Skyhawk	142083
Boeing	EB-29-50-BO	Superfortress	44-83965
North American	B-45A-1-NA	Tornado	47-063
Boeing	B-47E	Stratojet	50-064
Martin	B-57E	Canberra	55-4267
Douglas	RB-66B	Destroyer	53-452
Fairchild	C-119C-FA	Boxcar	50-142
Lockheed	C-130A-1-LM	Hercules	53-3131
Douglas	C-133A	Globemaster II	54-137
Douglas	D-558-2	Skyrocket	37974, NACA 144
Lockheed	F-80C-1-LO	Shooting Star	47-172
Republic	F-84F-40RE	Thunderjet	52-6560
North American	F-86H-1N	Saber	52-1980
Northrop	F-89D-NO	Scorpion	52-1830
North American	F-100C-1-NA	Super Saber	53-1716
McDonnell	JF-101A	Voodoo	53-2418
McDonnell	F-101A-5-MC	Voodoo	53-2430
Convair	F-102A	Delta Dagger	53-1796
Convair	TF-102A	Delta Dagger	54-1355
Lockheed	F-104A	Starfighter	55-2965
Douglas	F4D-1	Skyray	134919
Grumman	F11F-1F	Tiger	138647
North American	FJ-3	Fury	135792
Chance Vought	F8U-1	Crusader	140445
Vertol	H-21B	Workhorse	53-4380
Lockheed	T2V-1	Seastar	142535
Bell	X-1B		48-1385
Bell	X-5		50-1838
Chance Vought	XRSSM-N-9	Regulus II	GM-2007
Boeing	JB-17G-VE	Fortress	44-85813
Lockheed	C-130A	Hercules	54-6126

(Above, Center, Right) Bell X-1B 48-1385. Shown in NACA markings this is the third version of the X-1 family. It resulted from the second-generation contract of April 2, 1948, for four aircraft. Only three of these were built, the X-1A, 48-1384; X-1B, 48-1385; and X-1D, 48-1386. The X-1B arrived at Edwards on June 20, 1954, and was used by the Air force for pilot familiarization flights, being turned over to NACA on December 3, 1954. Its last four of seventeen NACA flights were made by Neil Armstrong. This was its last public showing at Edwards, and it was sent to the USAF Museum in January 1959.

(Right, Center) Bell X-5 50-1838. It was a real surprise to see the X-5 on display, this being the first of two built, making its first flight at Edwards on June 20, 1951. The second example, 50-1839, was lost during a spin test flight in October, 1953. The X-5 was the first variable sweep wing aircraft to fly, and the results aided the development of the much later F-111 family. Design of the X-5 was derived from the WWII German Messerschmitt P.1101, which was built, but never flown.

Douglas D-558-2. In contrast to the two Bell types, the D-558-2 Skyrocket was ordered by the Navy in January 1947, three being substituted for the last three D-558-1 Skystreaks. First flight was made at Edwards on February 4, 1948. Subsequently, it became the first aircraft to exceed both Mach 1 and then Mach 2. This is the second example, BuNo. 37974, NACA call sign 144, and the first one to be solely rocket powered, using a Navy version of the same motor which was used in the later Bell X-1s. It was handed over to NACA on August 31, 1951, and during the next 63 months it reached an altitude of 83,235 feet and a speed of Mach 2.005, making 105 flights before being retired on December 20, 1956, going to the National Air & Space Museum. Note its lack of air intakes for the Westinghouse J35 engine, originally fitted to the other two D-558-2s.

(Left, Center) Chance Vought XRSSM-N-9, GM-2007. This large ship-board surface-to-surface missile was somewhat similar to the USAF Northrop Snark, having a sweptwing, no horizontal tail, configuration. Fitted with a GE J79 engine in the production version, it was designed to cruise at Mach 2 after launch from a new class of very large Navy submarines. It had been kept very quiet, and had only been unveiled by the Navy at a conference in March 1957. One of the missiles was subsequently launched in a public demonstration at Edwards AFB during December 1957 (it was superseded by the new Navy Polaris missile). Some 57 feet long, it was fitted with a 115,000 lb. thrust solid rocket motor for launch. Its appearance at this open house was a complete surprise, proving to be its only appearance. It was finished in a smart looking insignia blue on upper surfaces and gloss white on lower surfaces.

(See also top p. 10) Boeing JB-17G-VE, 44-85813, (later N6694C). This B-17 was one of a few post-war engine test beds, having had its entire cockpit area moved to give more room for the test engine installation. Bailed to the Wright Aeronautical Company, the test engine seen is the big Wright R-3350 turbo-compound piston engine. Despite the testing, the engine was not ready for service with the airlines when installed into the Douglas DC-7 series, or the competing Lockheed 1049 Super Constellation, and caused the ultimate demise of Wright as a commercial engine manufacturer. The author had first hand experience of these problems when a TWA 1049 he was onboard had two successive engine failures on a trip from New York to London in September 1959, forcing unscheduled stops at Gander and Shannon.

North American B-45-1-NA 47-063. In the type's last display at Edwards, this B-45 was from the initial production batch, but with four 5,200 lb. thrust J-47 engines in place of the 4,000 lb. J-35 engines. It is shown with the Air Research Development Center (ARDC) flash on the tail, above a rather worn Arctic red marking (the Arctic red markings were large patches of insignia red paint applied normally to the aircraft's tail and wing tips, for help in finding the aircraft if it was forced down in Arctic snow conditions). This rather mundane approach to a jet bomber was soon superseded by the very advanced swept-wing B-47; the B-45's greatest claim to fame is that it was the first U.S. jet bomber. They also made the first covert over-flights of the Soviet Union, flown by RAF crews.

Martin B-57E, 55-4267. This American version of the English Electric Canberra was being used for temporary testing at Edwards, one of a batch of sixty-eight built as target-towing aircraft. Many of them were used for special tests, such as this example, and eventually were to be seen at Edwards for at least the next ten years.

Douglas RB-66B, 53-452. Operations in the Korean War had shown the USAF the need for a light bomber, to succeed the B-26. It was decided to convert the new Navy A3D Skywarrior design into a USAF type. However, the A3D had been designed at Douglas El Segundo, while it was decided to redesign and produce the new B-66 type at Long Beach. A contract was awarded in February 1952 for five pre-production RB-66As. Unfortunately, the small Douglas design team at Long Beach had very little experience at this type of work, and the program ended up as being declared one of the worst ever managed by the USAF. Part was due to the inexperience of the Douglas team, part due to the many design changes requested by the USAF, changing a virtual off-the shelf buy into a new design. Probably the biggest single mistake made in the redesign was to change the engine from the P & W J57 to the Allison J71, this causing the developed versions to be severely underpowered. First production deliveries began in February 1956, but eventually only 209 of all versions were produced. Incredibly, a second production line was set up atTulsa, this increasing costs unnecessarily. The displayed aircraft was from the largest production batch, of 145 built. It is fitted with a very large black panel just forward of the wing on the fuselage right side.

(Right, Center) Boeing EB-29-50-BO 44-83965. This special B-29 was used as a carrier aircraft; it is not listed as having been used in the X-1, etc., programs, but it has the bomb doors removed and large lifting shackles visible internally. It may have been a back-up aircraft, or else used on the Lockheed X-7 supersonic/hypersonic engine test vehicle program. It is in a very plain, natural metal finish, with an ARDC tail flash.

(Right, Top & Center p. 13) Douglas C-133A 54-137. The third production aircraft (there were no prototypes) of the fifty built; they played an important part in MATS (Military Air Transport Service) as being the only type able to carry outsize cargo, until the introduction of the C-5 in 1970. The type ran into flight control, stability, hydraulic, electrical, and fatigue problems, making its testing at Edwards rather prolonged.

Lockheed C-130A-1-LM 53-3131. The third C-130 of an initial 216 production batch; no one realized that the type would still be in production forty years later! It shows clearly the large bluff nose, sans the large radome fitted to later versions.

Chance Vought F8U-1 140445. Named the Crusader in Navy service, the F8U nose-mounted guns proved to be effective in the later Vietnam war. The unpainted surfaces on the second production aircraft clearly show the many different types of material used on the aircraft. Most unusual feature of the F8U was its variable-incidence wing, which rotated nose-up for take-off and landing. This was supposed to improve the landing characteristics of the aircraft by giving the pilot a better view from the cockpit on approach. However, it also considerably increased the drag of the aircraft to the point at which it was only too easy to get behind the "power-curve" and undershoot the landing point (this was because the drag of the aircraft increased faster than the power available from the engine, even in afterburning). Well-trained Navy pilots avoided this situation, and the F8U went on to score splendidly in the Vietnam war. A later version of the aircraft with the same feature was the XF8U-3, a truly Mach 2 plus fighter, but it lost out to the McDonnell F4H-1 in testing at Edwards. Because the tests ended some time before that years open house, the XF8U-3 was never seen publicly (this was true for several types tested at Edwards).

Lockheed T2V-1 142535. Fourteenth Seastar of the 150 built, the type was rarely seen at Edwards. It entered service at NAS Pensacola in late 1957, but maintenance problems with its boundary-layer control system caused it to be withdrawn in favor of the North American T2J-1 circa 1960 (however, several survived as unit "hack" aircraft for many years). As such, it is one of Lockheed's least known types. It is finished in the then usual Navy trainer white and red finish.

Douglas A3D-2, 138905. Nicknamed the "Whale" due to its size, the A3D served the Navy for almost forty years. Seen is the third of the first 123 A3D-2s; it had a stronger airframe and more powerful J-57 engines. It is finished in the then current Navy paint scheme of dark gloss sea blue above and below, with white areas on the nose radome, vertical tail, and wing tips, plus trim stripes on the engine cowls and fuselage speed brakes. Designed by the legendary Ed Heinemann at El Segundo, the A3D came in at a gross weight of 70,000 lb. versus the Navy specification requirement of 100,000 lb., a truly extraordinary achievement. Although designed as the Navy's carrier strategic nuclear weapon system, that role disappeared in a few years, and the type soldiered on in a variety of key services. Little known is that Pan American Airways wanted to buy two A3Ds in 1957 for use in training crews for their forth-coming civil jet transports. The deal fell through because Douglas could not deliver them in time, due to very heavy Navy commitments.

(See also Top p. 16) Grumman F11F-1F, 138647. Last of the single-engine Grumman fighters, the F-11F-1 did not stay in service very long. It entered service in March 1957, with VA-156, but recurring problems with its Wright J65 engine caused it to be relegated to the training role in 1959. The F11F-1F version had the General Electric J79 engine fitted, giving the aircraft Mach 2 performance, but its short range rendered it surplus to Navy operational requirements. The F11F-1 is chiefly remembered for its performance with the Navy Blue Angels.

(Center, Right) Douglas A4D-2, 142083. Production of this version of the Skyhawk totaled 542. This is the second aircraft and shows the new refueling probe and the single-surface, externally stiffened rudder. Unlike the F11F-1, although fitted with the same Wright J65 engine, the type was very successful in service. Perhaps the fitting of a refueling probe on the right side of the A4D's nose rectified the chief problem attributed to the F11F-1, its short range.

McDonnell JF-101A, 53-2418. This is the first F-101 built, having made its first flight on September 29, 1954. Originally designed for SAC, it was turned over to TAC before its first flight. Second of the USAF Century series, it suffered from pitch-up control problems, but more than 800 were built, including those used by the Royal Canadian Air Force.

(Center, Left, see also top p. 18) McDonnell F-101A-5-MC, 53-2430. This was the thirteenth F-101A built: compare with the prototype above. The ninth aircraft, 53-2426, was bailed to Pratt & Whitney for development of the more powerful J57-P-55s that were going to be used in the F-101B version. This combination set a new speed record of 1,207.6 mph in December 1957.

(Right, Bottom) Convair TF-102A, 54-1355. Fifth of 63 TF-102As built, this type provided essential training for the single-seat interceptor version. Its bulky appearance is due to the side-by-side seating for the instructor and pupil.

Northrop F-89D-NO, 52-1830. Second of the eighth batch of F-89Ds, this is seen with the new Douglas MB-1 Genie unguided nuclear rockets under the wings. The large main landing gear wheels were nicknamed "Santa Fe wheels," as they looked like the large driving wheels used on the railroad's steam engines!

Republic F84F-40RE, 52-6560. The prototype YF-84F (originally the YF-96A) first flew in June 1950, so it was well into service when seen at Edwards. Major feature of the F-84F was the fitting of an all-flying horizontal stabilizer, giving the aircraft much better trim control at high subsonic and transonic speeds. F84Fs were used by regular USAF units until 1964; they also served with many ANG units until 1971, resulting in a long service career.

North American F-86H-1N, 52-1980. Similarly to the F-84F, the F-86H served with many units starting in 1954 (473 were built), but they were all replaced by F-100s by mid-1958, also being transferred to ANG. At Edwards, they were mainly used by the test-pilot school, and for chase duties. Only four years earlier, the F-86 had established a remarkable record in the Korean War, and it was still held in high esteem at this time.

Vertol H-21B, 53-4380. 163 H-21Bs were built for the USAF, and this is the 130th. Normally used as an assault transport by Troop Carrier Command, they were used in rescue work at Edwards. Vertol was the chief developer of the twin fore-and-aft rotor concept, and it is still in use at the end of the century, in the form of the CH-47 Chinook family.

(Center) Lockheed F-80C-1-LO, 47-172. This was the second production aircraft of the F-80C, last version of USAF's first operational jet fighter. Powered by an Allison built GE J33 centrifugal compressor development of the original British de Havilland H-1 engine, it always seemed to be underpowered, requiring long takeoff runs. This was the last time we saw a F-80 at Edwards; the few remaining serving in ANG and AF Reserve units.

North American F-100C-1-NA, 53-1716. The eighth production F-100C of some 476 of this version built, the type was the main USAF tactical fighter at this time; the later F-100D version having made its first flight in January 1956. Being the first of the so-called Century fighter series, it had a marked influence on its contemporaries at home and overseas. Apart from running into stability and control problems, from having had the original height of its vertical tail reduced (to save weight), the F-100 had more mundane problems. For example, the drag chute compartment had been located on the bottom of the aft fuselage, requiring flexible steel tabs to cover the run of the chute riser line up to its attachment below the rudder. The problem with this installation was that all fluids and rain water tended to drain into the compartment. After flying at high operational altitudes, when the aircraft deployed its drag chute on landing, it often came out as a solid, frozen block! This resulted in a much longer than desired landing run, leading eventually to the installation of back-up arresting hooks on these land based fighters. Moreover, in England when the English Electric Lightning was being developed, the designers copied the F-100 chute installation completely, thinking that North American knew what they were doing! (from a conversation with the Lightning chute system designer, some years later).

(Left, Center) Lockheed F-104A, 55-2965. Eleventh of a batch of fifteen YF-104A test aircraft, this shows the new GE J79 engine, with fixed flow control cones in the engine intakes, a lengthened fuselage with forward retracting (instead of aft retracting) nose, and main landing gears. One of this batch of aircraft reached Mach 2 in April 1955, but delivery of operational F-104As did not begin until January 26, 1958. The test program proved to be long and frustrating, and the type lost the most aircraft of any of the Century series during this testing. However, USAF and Lockheed determination to make some recovery on their investment in the F-104 was to lead to much bigger things (such as the F-104G export version), of which, more later.

Douglas F4D-1, 134919. Known as the "Ford," this example is from the second main production batch, 420 eventually being produced by December 1958. The Skyray's main problem in service was its huge access door under the belly. Completely exposing the engine when removed, this door and its attachments did not stay aligned after a few carrier deck landings, becoming increasingly difficult to reinstall. Thus, the type was eventually relegated to land-based units. It had a very good performance with the after-burning P & W J57 engine, but it was the second prototype XF4D-1 with the original Westinghouse XJ40-WE-8 engine that brought the world aircraft speed record back, from the British Supermarine Swift Mk. 4 (which the author had worked on).

(Left, Center) Lockheed C-130A, 54-6126. This is the fifteenth C-130 built, with no visible external differences from the earlier ones, except for the addition of the Arctic red panel on the vertical tail.

Fairchild C-119C-FA 50-142. Developed from the original twin-boom C-82 transport, this example is from the second production batch, following the 54 C-119As. It introduced the long dorsal fins, and deleted the horizontal stabilizer sections outboard of the booms. Used widely in Korea, the type could carry 62 troops. Later, special versions were used to retrieve USAF satellites (such as the Samos) re-entering over the Pacific Ocean.

(Left, Center) Douglas B-66B, 54-450. This Destroyer is also carrying the Arctic red markings, but unlike the C-130, it is also carrying bands of it across the wings. Note the dielectric pods at the wing-tips (not fitted to the static display RB-66B). It makes an interesting comparison with the A3D Skywarrior taxiing in with the wings folded.

McDonnell F3H (serial number not clearly visible). The Demon turned out to be the least successful of all of the production McDonnell fighters, chiefly due to being designed around the abortive Westinghouse J40 engine (like many of its contemporaries). A change to the Allison J71 engine saved the Demon from extinction, but production ended in December 1959, as the type was severely underpowered. Remaining in service until 1965, the Demon was also McDonnell's first production sweptwing type (their first sweptwing design to be built was the XF-85), and efforts to rectify its shortcomings led directly to the F-4 Phantom II.

Northrop F-89C, 51-5802. Some 164 of this third version of the Scorpion were built, retaining the original six 20mm cannon armament in the nose, and adding fins to the wing tip tanks. The tanks are colorfully decorated in red, with white diamonds on this example. Scorpions were rarely seen publicly at this time, being the main type with Air Defense Command.

McDonnell F-101A, 54-1440. This is the 32nd Voodoo of the 77 of the "A" model to be built. First use of the F-101A was by the 27th Tactical Fighter Wing, in May 1957, just a few days before this display. This was the second Century series fighter, and its worst feature was the adoption of the German-inspired T-tail. This horizontal tail position was very bad for maneuverable aircraft, due to sudden pitch-up control problems. This was caused by the wing downwash suddenly impinging on the horizontal tail at high angles of attack of the aircraft. It was found that the best position for the horizontal tail in such aircraft was for it to be as low as possible (as on the F-100, for example). The only cure for the F-101's pitch-up was the fitting of an inhibitor, restricting aft stick movement (this, of course, restricted the maneuverability of the aircraft). McDonnell did not use the T-tail on any of their subsequent fighter designs. The later F-101C was able to carry a tactical nuclear weapon on the center fuselage pylon, and had a beefed-up wing for low-level operations.

Douglas A3D (serial number not clearly visible). Seen taxiing back after its flight demonstration, this "Whale" is seen with both the wings and vertical tail folded.

Boeing B-47E, 52-221. This is the 29th B-47E to be built at the huge Air Force plant in Marietta, Georgia, operated by Lockheed. The production B-47Es followed eight B-47B models, first to be built at Marietta, which was the third source for SAC.

1958

Having learned from my first visit that it was essential to get to Edwards early if I wanted good, unobstructed photographs, I stayed at Lancaster and followed the details described in the introduction. This paid off handsomely, and in addition the weather was superb, bright and clear. This also resulted in a lot of contrails in the vivid blue sky, clearly visible in this year's photographs. Among the new types this year were the Boeing KC-135A, North American F-100F, Lockheed F-104B, Republic F-105B, Convair F-106A, North American XF-107, Bell X-1E, Sikorsky H-37, and the Lockheed Jetstar. The new Convair B-58 was also present, but only in the dark hangar. There were also some older aircraft that put in an appearance, including a North American F-51D, Douglas C-47A, and a Douglas C-54A. The brilliant light also brought out the tremendous polish on many of the displayed aircraft. A few of them were also showing off the use of a vivid fluorescent red paint, a foretaste of things to come. The listing of the aircraft present follows.

Manufacturer	Type	Popular Name	Aircraft Serial
Douglas	A3D-2	Skywarrior	142243
Douglas	A4D-2	Skyhawk	142760
Boeing	JB-17G-VE	Fortress	N6694C
Boeing	JB-47B-26BW	Stratojet	51-2073
Martin	B-57A	Canberra	52-1419
Convair	YB/RB-58A	Hustler	55-665
Douglas	RB-66B	Destroyer	53-428
Douglas	C-47A	Skytrain	43-48188
Douglas	JZC-54A-DO	Skymaster	41-37268
Fairchild	C-119C-FA	Flying Boxcar	51-2586
Fairchild	C-123J	Provider	54-647
Lockheed	C-130A	Hercules	54-1626
Douglas	C-133A	Globemaster II	54-135
Douglas	C-133A	Globemaster II	54-137
Boeing	KC-135A-02-BN	Stratotanker	55-3124
Douglas	D-558-2	Skyrocket	NACA 145
North American	F-51D	Mustang	NACA 148
Republic	NF-84F-25RE	Thunderjet	51-1621
North American	TF-86F-NA	Saber	53-1228
North American	F-86F-NH	Saber	51-13171 et al.
North American	F-86H-1N	Saber	52-1980
North American	JF-100C-1-NA	Super Saber	54-2114
North American	F-100F-1-NA	Super Saber	56-3726
McDonnell	JF-101A-5-MC	Voodoo	53-2426
Lockheed	F-104A	Starfighter	55-2971
Lockheed	F-104B-5-LO	Starfighter	57-1297
Republic	F-105B-RE	Thunderchief	54-0100
Convair	JF-106A-CO	Delta Dart	56-0453
Convair	F-106A-CO	Delta Dart	56-0462
North American	XF-107A		NACA 207
Douglas	F4D-1	Skyray	134853
Grumman	F11F-1F	Tiger	138646
Chance Vought	F8U-1	Crusader	141339
Hiller	YH-32	Hornet	55-4971
Sikorsky	H-37A-SI	Mojave	54-998
Cessna	YH-41-CE	Seneca	56-4237
Lockheed	T-33A-5-LO	Shooting Star	57-567
Bell	X-1E	—	NACA 6063
Lockheed	CL-329	Jetstar	N329J

Sikorsky H-37A 54-998. This very large helicopter was originally developed for the Navy/Marines as the HR2S-1, first flying in 1953. It was the first Sikorsky design to use two engines. The Army tested one as the YH-37 in 1954 and eventually ordered a total of 94. This is the 6th one, undergoing testing at Edwards. Capable of holding 36 troops, the last one was delivered in May 1960.

North American TF-86F 53-1228. The two-seat trainer version of the F-86F was the second one built, making its first flight in August 1954. The fuselage was lengthened by 63 inches and the wing moved forward 8 inches; this version had a taller vertical tail with a ventral fin, and included two nose guns. However, by this time the USAF had decided that they did not need the type, as the new F-100 was already in production to replace the F-86s.

Douglas B-66B-DL 53-428. This is the first of the only bomber version of the Destroyer, and 72 were built at the Douglas Long Beach plant. It could carry up to 15,000 lb. of bombs. Thirteen of them were converted to EB-66B-DLs as Electronic Warfare radar-jammers, for use early in the Vietnam war.

Bell X-1E 6063. This version of the X-1 family became necessary after the loss of the X-1D and the third X-1 aircraft. Both of them were lost before they could enter the test program, so it was decided to modify the second X-1 into the X-1E. This included the incredibly thin, 4% wing, a new engine turbo-pump, and an entirely new canopy. It also had two ventral fins fitted to improve directional stability at speeds over Mach 1.6. They markedly improved the stability, but also decreased the aircraft's maneuverability over the entire speed range. The X-1E made its first flight in December 1955, and ended up making 26 flights before being retired. During these it reached a speed of Mach 2.24. It can be seen to this day, mounted on a pylon at NASA's Dryden Research Center, Edwards AFB. The name "Little Joe" on the left side forward nose referred to the test pilot, Joe Walker, who made the first twenty-one flights in the X-1E. The yellow pair of dice, showing an "eleven," apparently referred to Walker's liking for their use. Walker was the NACA/NASA test pilot killed in the unfortunate mid-air collision between his F-104 and the second XB-70 (see the 1966 chapter for more on the XB70A-2).

(Right, Bottom) North American XF-107A NACA 207. This unusual looking fighter was originally designed as the F-100B, but its drastic configuration changes caused the re-designation. It was entered into competition with the Republic F-105, but lost out in March 1957. Three were built, being turned over to NACA for a while.

(Right, Center) Convair JF-106A-CO 56-0453. Seen is the third F-106 built, the prototype making its first flight in December 1956. It was essentially a very cleaned up version of the F-102, fitted with the more powerful J75 engine. The F-106A first went into service with the 489th Fighter Interception Squadron at Geiger AFB, Washington, in 1959, and remained in service until 1988. Remaining examples were then converted to QF-106 drones.

(See also Top p. 29) Lockheed F-104A 55-2971. This example was the last of the service test batch of 15 YF-104As. The type first reached Mach 2 in April 1955, and deliveries of the production F-104A were made to the 83rd Fighter Interceptor Squadron at Hamilton AFB in January 1958 (four months before this display). Just before the display, a F-104A had set an altitude record of 91,249 feet on May 7, and a speed record of 1,404 mph on May 16 (day before the display). Despite these records, the F-104A only remained in service with Air Defense Command until 1960, being withdrawn for a while, returning from service with ANG for a time, and then being added to by F-104Cs from TAC.

(Center, Bottom) Republic F-105B-1 54-0100. Here is the third F-105, the first J75 powered version. It made its first flight in May 1957, and beat out the North American F-107 for adoption as a supersonic tactical fighter-bomber, as related above. 78 F-105Bs were built, and the first service versions went to the 4th TFW in May 1958 (just after this display). The very high cost of the F-105Bs eventually caused production of the later F-105D to be terminated in favor of the cheaper McDonnell F-4C.

(Left, Center) Lockheed Jetstar N329J. This was a real surprise, a civilian twin-jet transport from Lockheed. It resulted from USAF 1956 requirements for a UTX utility transport, which asked for industry-funded prototypes. The Jetstar made its first flight in September 1957, fitted with two British Bristol Orpheus engines. The original plan for this engine to be produced by Wright fell through, and Lockheed had to convert the design to a four-jet type, with Pratt & Whitney J60 jets. In that form, five were originally ordered by the Air Force in 1960, as the C-140A.

(See also top p. 32) McDonnell JF-101A-5-MC 53-2426. Seen is the 9th F-101A, being used for special tests at Edwards. It has a certain amount of red paint under the horizontal stabilizer, and the forward fuselage; compare it with last year's version.

Douglas F4D-1 134853. Another of the Navy's new production fighters, this F4D-1 was the 109th aircraft from the second production batch, also in the new gray and white paint scheme.

Douglas C-47A 43-48188. This old-timer was being used as an ARDC staff transport at this time, immaculately maintained.

North American F-100F-1-NA 56-3726. Second of the F-100F two-seat trainer versions, this was fully equipped for operational use, but only carried two guns in the nose. The type made its first flight in March 1957, some 339 eventually being built. April 1958 (just before this display) saw the beginning of the transfer of the original F-100As to ANG use.

North American F-100C-25-NA 54-2114. Sixth from last of the El Segundo built F-100Cs, being succeeded by the F-100D in early 1956. This one is being used by the then Air Research & Development Center (ARDC), note the vertical tail marking.

North American F-86H-1N 52-1980. Fourth of the production F-86Hs, this is also in ARDC service, with a bright red band around its nose.

(Center, Bottom) North American F-51D-25-NT NACA 148 (ex 44-84958). We certainly did not expect to see a F-51D in use at Edwards, or on display, some fifteen years after its first flight. This one was now being used by NACA; note the colorful four-eyed green character "CEECE," (from a then-current TV series) painted on the nose cowling (significance not known). This particular batch of F-51Ds contained four two-seat versions, and this may be one of them, apparently being used for pilot familiarization at Edwards.

(Top, Center, Bottom) Republic NF-84F-25RE 51-1621. This is a very unusual version of the F-84F, being extensively modified to test an early jet thrust reverser designed by Rohr. Designed to reduce the landing roll of the aircraft, the complications of the reverser outweighed the service use of re-usable drag parachutes, and it did not go into use for that purpose. Note the faded dayglo red paint applied to various areas of the airframe, a portent of more to come.

Lockheed T-33A-5-LO 57-567. Production of the T-33 continued until 1959, and this is the 37th one from the 1957 order, some 474 aircraft from the last of 5,691 built; 699 were for the Navy and 1,058 for the Mutual Aid Program (MAP). The T-33 had a very long life, having first flown in 1948. This one also carries the ARDC markings, including the bright red decorations.

(Center, Left) Chance Vought F8U-1 141339. This is the third aircraft from the second production run of the successful Crusader, widely heralded as the last Navy fighter with guns. As distinct from the prototype we saw earlier, this one is finished in the 1956-adopted Navy light gull gray and white camouflage scheme, replacing the earlier overall gloss dark sea blue.

Grumman F11F-1F 138646. One of two Tigers built with the GE J79 engine to gain an all-round performance improvement; major efforts were made to sell this version to the Japanese, but to no avail. The test aircraft color scheme is in white with red trim on nose and vertical tail.

(Center, Right) Boeing JB-47B-26BW 51-2075. Retained at Edwards for continuing test work, this the 30th aircraft from the last batch of 399 B-47Bs built. Very anonymous in natural metal finish, the B-47 was SAC's main jet bomber at this time.

(Right, Center) Douglas A4D-2 142760. 542 of this version of the Skyhawk were built, this being the 86th from the third production run of 280 aircraft. It is seen with the large under-belly refueling pod.

Martin B-57A 52-1419. This unassuming looking natural metal finished B-57 is in fact the second to be built, retaining the original British side-by-side cockpit seating. Only eight of these were built, followed by a batch of 67 RB-57As, before the change was made to a tandem cockpit in the B-57B version. It should be recalled that the B-57 is the only USAF combat aircraft of foreign design to be built since the end of WWII.

(Top, Center) Douglas A3D-2 142243. The A3D Skywarrior was now in service with the Navy, 164 of this version being built, this being the 7th from the second production batch. It was to be the definitive production version, entering service with eight heavy attack squadrons on the Navy Essex and Midway class aircraft carriers. Many of them were later modified as VA-3B personnel carriers, KA-3B tankers, and EKA-3B ECM/tankers. This one is in the new gray and white finish, with red conspicuous bands on the wing and tail surfaces. Note the unusual use of red for the large number on the nose, more normally in black.

(Bottom, See also p. 40) Boeing KC-135A-02-BN 55-3124. This was the first time the new KC-135 Stratotanker had been displayed at Edwards, causing much interest, not least due to its relationship to Boeing's new commercial jet transport, the 707. This is the sixth of the initial production version, seen with the original short vertical tail. It is not generally known that this is the Boeing Model 717-100A, filling in the missing gap in the 7-7 series. Note that the fuselage does not have a crease at the floor level, unlike all of the commercial 707s: the 707s have an upper lobe four inches larger in diameter, required to squeeze in an extra row of seats. This was at Pan American's request after Douglas had offered them the DC-8 with six-abreast seating. Boeing had no option but to add the extra space, thus ruining their plans to produce the KC-135 and 707 on the same production lines. It is interesting to note that this basic fuselage cross-section has been used in the Boeing 707, 727, 737, and 757 aircraft, surely a record for any type of aircraft. This aircraft had a very long life, making its first flight in April 1957, but not being attrited until March 1991. During those thirty-four years, it was successively a test KC-135, a JKC-135A, a NKC-135A, and finally a GNKC-135 (being primarily a test aircraft for most of its life).

U.S. AIR FORCE
NAVY

53124

53124
NAVY

(Center, Right) Douglas C-133A 54-137. Third of the 35 C-133As, the Cargomaster shows off its bulk, next to the C-135. Finished with Arctic red markings, the type suffered from numerous problems in service, but its huge freight hold allowed all of the operational USAF IRBMs and ICBMs (Atlas, Jupiter, Thor, and Titan) to be carried. A larger dorsal fin was fitted after flight testing. Incidentally, a C-133A fuselage was in a water-test tank outside the Douglas Long Beach engineering offices for several years, indicative of the type's fatigue problems (particularly in the fuselage area inboard of the propeller discs).

(Right, Center) Boeing JB-17G N6694C. Originally 44-85813, built at the Wichita plant, this was the first B-17 to be converted for use as an engine test in 1946, and was subsequently turned over to the Wright Aeronautical Co. It is seen with a Wright 3350 turbo-compound engine in the nose. This had replaced the earlier Wright 5,500 hp XT-35 Typhoon turboprop engine.

(Left, See also bottom p. 42) Fairchild (Chase) C-123J 54-647. This interesting design stemmed from the Chase XG-20 cargo glider built in 1949. When fitted with two piston engines it became the XC-123, first flying in October 1949. Problems with later production efforts led to the type being handed over to Fairchild. The first Fairchild C-123 Provider (54-552) flew in September 1954. After fitting of the large dorsal fin, some 302 were built. This version was the 94th one built by Fairchild, and shows an interesting landing gear designed for use on snow. Also fitted are two wing tip Fairchild J44 jet engines, this version being designed to serve as a DEW-Line (Distant Early Warning radar system) support aircraft. Ten were built and flown by the Alaskan ANG. The C-123 must be the only transport type flown as a glider, piston-engine, piston plus jet, and four-jet transport! There was once a fanciful story that the design of the later C-130 had been based on a proposed four-engine version of the C-123, but the C-130, C-132, and C-133 all came from a top level USAF decision in 1951 to equip with turboprop transports, long before the C-123 entered production. However, there is no doubt that it led the way in the military transport field, with its large rear loading ramp.

(Center, Left) Douglas JZC-54A-DO 41-37268. A super clean example of the WWII C-54, in use by ARDC. It has been fitted with updated radar grafted onto the original nose, possibly hence the fancy JZC designation.

(Center, Bottom, Top p. 45) Douglas C-133A 54-135. Flying over and then taxiing past, the first C-133 shows off its bulk and very long span tapered wing. The root to tip taper ratio on this wing was in excess of the maximum four-to-one ratio found to be the largest giving good wing stalling characteristics (from Royal Aircraft Establishment tests at Farnborough, England, before WWII). The C-133A later added many vortex generators in front of the ailerons, in an attempt to improve roll control near the stall.

North American F-86F-NH, 51-13171, et al. These were the standard chase aircraft used by ARDC at Edwards at this time, carrying red under-wing drop tanks. They were also used to buzz the Antelope Valley residents to let everyone know that Edwards was having its open house!

(See also Top p. 46) Fairchild C-119C-FA 51-2586. Seen dropping some heavy cargo is a stock C-119C, of which 303 were built by Fairchild (including 41 by Kaiser at the old Ford Willow Run plant).

Convair F-106A 56-0462. This vividly red-painted aircraft is the eleventh production Delta Dart.

Lockheed C-130A 54-1626. Another eleventh production aircraft is this C-130A, taxiing by with the outboard engines shut down. Only the first 27 C-130As did not have the radar in the nose; some of these were later fitted with it, changing the nose appearance drastically.

1959

The third visit to Edwards showed several new aircraft shining on a nice clear day; several of these were to become very well known, while others were only to be seen this year. The new types outside included the huge Boeing NB-52A, carrier for the new North American X-15, McDonnell F4H-1, a trio of experimental VTOL types—the Bell XV-3, Doak VZ-4, and the Vertol VZ-2. A new liaison type, the Helio L-28, was seen in a fly-past. We were also treated to two very different commercial four-jet transports, the large Douglas DC-8 and the surprising McDonnell Model 119. Tucked away in the dark hangar were the new Northrop T-38 and the spin test Republic F-105. Of the new types, the three VTOL types were never seen again, the B-52/X-15 combination went on for years, the McDonnell F4H-1 became a prime USAF and U.S. Navy type, the Douglas DC-8 became a famous transport, and the McDonnell 119 faded into obscurity. It was very noticeable this year how many aircraft were finished with large areas of brilliant fluorescent orange and red paint, intended to try and improve visibility of aircraft flying near Edwards to help prevent collisions. In retrospect, a good year. The list of aircraft seen follows.

Manufacturer	Type	Popular Name	Aircraft Serial
Douglas	A3D-2	Skywarrior	142243
Douglas	A4D-2	Skyhawk	142083
Boeing	JB-47B-26BW	Stratojet	51-2075
Boeing	NB-52A-BO-1	Stratofortress	52-003
Boeing	B-52D-BW	Stratofortress	not visible
Martin	B-57B	Canberra	53-3864
Douglas	RB-66A (CJ805)	Destroyer	52-2828
Douglas	JRB-66A	Destroyer	52-2830
Douglas	C-54A-DO	Skymaster	41-37268
Douglas	C-133A	Globemaster II	54-135
Boeing	KC-135A	Stratotanker	not visible
North American	F-86F-25-NH	Saber	52-5412
North American	F-100D-25	Super Saber	55-3681
North American	F-100F-1-NA	Super Saber	56-3726
McDonnell	NF-101A	Voodoo	53-2418
Convair	F-102A (J85)	Delta Dart	54-1398
Convair	TF-102A	Delta Dart	54-1354
Lockheed	F-104A	Starfighter	55-2966
Republic	JF-105B-RE	Thunderchief	54-0100
Republic	F-105B-RE	Thunderchief	54-0102
Convair	JF-106A-CO	Delta Dagger	57-237
North American	F-107-1-NA		55-5120
McDonnell	F4H-1	Phantom II	142259
Chance Vought	F8U-1M	Crusader	141340
Vertol	H-21	Workhorse	not visible
Helio	L-28A-HE	Super Courier	58-7025
Lockheed	T-33A-5-LO	Shooting Star	57-562
Northrop	YT-38	Talon	58-1192
Bell	XV-3		54-148
Vertol	VZ-2		56-6943
Bell	X-1E		NACA 6063
North American	X-15		56-6671
Doak	VZ-4		56-9642
Douglas	DC-8-11		N8033D
Lockheed	CL-329	Jetstar	N329J
McDonnell	Model 119		N119M

(See also Top, Center p. 49) Northrop YT-38 58-1192. The Talon was the first supersonic trainer developed for the USAF. This is the second prototype, seen in the giant hangar at Edwards while being prepared for its first flight, made on June 12, 1959. Note that the rear cockpit cover has been painted over, to help protect test instrumentation from the intense glare at Edwards. No less than 1,189 T-38s were eventually built, and they are currently being reworked to serve long into the next century. The author worked on the design of the main landing gear, then became responsible for the spin recovery system (see Appendix 3).

Republic F-105B-1 54-0102. Seen is the third F-105B, 78 of which were built. This version had possibly the sharpest nose of any USAF fighter. First of the unit aircraft had been delivered just a year earlier (three years late!), to the 335th TFS, of the 4th TFW, then at Eglin AFB, and later at Seymour Johnson AFB. Republic was having serious problems integrating the aircraft's complex subsystems, causing major production delays. Thus, the unit did not reach full strength until mid-1959. Consequently, in March 1959, the Air Force canceled production of the F-105E, which was the second proposed two-seat version of the F-105.

(Left, Center) Republic JF-105B-1 54-0100. Prototype of the F-105B series, this had made its first flight some two years previously. It is seen in modified form, as the spin test aircraft. Note the extensive external stiffening added to the rear fuselage; this was to take the very heavy loads imposed by the large recovery parachute housed at the aft end, behind the rudder. To the best of the author's knowledge, this was the only aircraft at Edwards ever finished with fluorescent yellow paint.

Bell X-1E NACA 6063. This was the last year we were to see one of the pioneering X-1 series, in the last version of the X-1E. It is fitted with its final ventral fins for increased stability control, now painted white. The rudder, however, remained in natural metal finish.

Bell XV-3 54-148. 1959 became noticeable for a number of new V/STOL prototypes shown. This is the Bell tilt-rotor entrant, the second of two built. Powered by a single 400 hp R-985 piston engine in the center fuselage, it was built for Army evaluation, originally designated as the XH-33.

Doak VZ-4 56-9642. Another new Army test vehicle, the unusual Doak VZ-4 was the sole prototype. It was powered by a 1,000 hp T53 turboprop, driving two tilt-able ducted fans. Unlike the VZ-3, it was finished in a smart gloss olive drab, white, and red color scheme.

Vertol VZ-2 56-6943. Yet another of the new breed was the Vertol model 76, a tilt-wing vehicle. Built with Army funds, it was tested by them, the Navy, and NASA. It could be considered the beginning of the long trail leading to the current Bell/Boeing V-22 Osprey. Like the VZ-2, it was finished in natural metal, but with large areas of fluorescent red.

North American F-107-1-NA NASA 55-5120. Here is the third prototype XF-107, finished similarly to the previous year NACA version, but with NASA on a plain yellow band across the vertical tail, together with the USAF serial number. Previously, the aircraft had had the NACA number "207" on the tail, and a large white "XF-107A" marking on the nose. NASA had recently been formed out of the old NACA when the Soviet Sputnik space satellite gave the USA a huge technological shock.

Convair JF-106A-CO 57-237. This was the 25th F-106A built, and is seen in the standard gloss aircraft light gray finish. Only visible difference to normal production aircraft is the white letter "A" at the top of the vertical tail, but many of these early aircraft were used in continual upgrade programs.

Convair TF-102A 54-1354. The Delta Dagger was seen on display at Edwards for many years; this year's had the new ARDC tail markings.

Lockheed F-104A 55-2966. Seen is the 11th of the prototype YF-104A batch, which became F-104A-1-LOs. This was the last year that aircraft from this batch were to be seen publicly at Edwards. Note the extensive areas painted with fluorescent orange paint. This year, the Air Force was using lots of fluorescent paint as an anti-collision preventative and used both fluorescent orange and red finishes, applied over a white paint base. However, they soon discovered that both finishes faded quickly, and that the red eventually faded to orange. Thus, they dropped the use of the orange and standardized on the red for a while.

(Top, Left) Convair F-102A (J-85) 54-1398. Here is an interesting test bed version of the F-102A, modified to carry a GE J-85 engine mounted in a retractable under belly pod (note how the aircraft has had to be mounted on blocks to allow the pod to be shown extended). It is finished with lots of fluorescent red over the normal aircraft gray.

North American F-100D-25NA 55-3681. This year was the first time we saw the final single-seat version of the F-100 series, the F-100D. It was built in large numbers, this being the 362nd one. Note how the fluorescent orange finish is flaking off the natural metal surfaces of the aircraft.

North American F-100F-1-NA 56-3726. Seen is the second of the 339 F-100F two-seat version of the Super Sabre, retained at Edwards for use by ARDC. A lot more of the aircraft's surface is covered in fluorescent red than on the adjacent F-100D.

North American F-86F-25-NH 52-5412. Still in use as chase aircraft, the F-86Fs also succumbed to the pervasive fluorescent paint, but this ARDC one is fading badly, due to lots of exposure. However, its under-wing drop tanks are finished in regular red and yellow paint.

Lockheed T-33A-5-LO 57-562. The perennial T-33 is seen with only minimal fluorescent paint applied, on the tip tanks, and an incomplete ARDC tail marking.

(Left, Center) McDonnell NF-101A 53-2418. The prototype was now bailed to GE for use as an engine test bed, for the J-79 engine development. It is seen resplendent in a fresh coat of fluorescent red, with the GE logo on the tail. Note that GE was the only engine manufacturer to have their development center located at Edwards.

(See also p. 60) McDonnell F4H-1 142259. One of the most important new types to appear at Edwards this year was the huge McDonnell F4H-1, powered by two J79-GE-3 engines on loan from the USAF. Originally designated as the AH-1, the type was re-designated as the F4H-1 prior to contract award. One unfortunate result of this change was the removal of the originally planned cannon armament in favor of an all-missile armament. This is the first of two prototypes, with fluorescent red over the Navy light gray and white finish. The author had worked on the design of many portions of this aircraft, including the wing fold mechanism, and the arresting hook. He was also responsible for the design of the dual-parachute spin recovery system used on the test aircraft (see Appendix 2). It is not generally known that the entire wing and its systems, and the aft fuselage and empennage were designed at Northrop under a sub-contract from McDonnell, as they were then overloaded with design work on other aircraft. During this design period at Northrop it was found necessary to make some major wing and empennage configuration changes. It was found that the flat (no-dihedral) wing should have had three degrees of dihedral from root to tip to provide sufficient stability. Unfortunately, the very solid wing jigs were already built and sitting on the shop floor, waiting for the first aircraft parts. Some clever designer (unfortunately, unknown) came up with the idea of bending the wing up some twelve degrees at the wing-fold joint, providing the same effective wing dihedral. At about the same time, wind tunnel tests showed that the flat horizontal tail was being impinged by wing down-wash at high angles of attack, causing sudden un-commanded pitch-up problems. It was found necessary to bend the horizontal stabilizer down to get it out of this wash, because it could not be lowered any further on the small aft fuselage. At first, it was bent down about sixteen degrees, but this did not prove sufficient to provide a cure, so it was bent down to slightly more than twenty-three degrees to obtain the desired results. Thus, arose the famous Phantom II bent up-down appearance. Note the original nose radome, cockpit, and engine intakes used on the prototypes; these were subsequently enlarged on later production versions.

NAVY
142259
NAVY

F4H-1

Boeing JB-47B-26BW 51-2075. Spared from the fluorescent paint onslaught, this B-47 looks positively unfinished, but it is one of the test aircraft retained at Edwards for continuing improvements to SAC's then main jet bomber (but B-52s were by then beginning to supplement the B-47s).

Chance Vought F8U-1M 141340. The tenth production Crusader had a white fuselage (except for the aft portion), and fluorescent red surfaces. The white finish was then coming into vogue on test aircraft, as it made it easy to check for fluid leaks, smoke, and other problems (also see the YT-38A).

Douglas RB-66A, 52-2830. This is the third RB-66A to be built, intended for all-weather night photographic reconnaissance use. It is pristine in its natural metal finish, with not a hint of any extra colors or markings. The B-66 was to perform sterling service in the forthcoming Vietnam conflict, not being retired until the early 1970s.

(Top, Right) Douglas A4D-2 142083. Here is the second A4D-2, as seen in 1957, still in its highly polished natural metal finish, but this time it has been extensively painted with fluorescent red. However, the rudder and the drop tanks are in white paint finish. There is an interesting crossed fingers marking painted on the nose (regrettably, we don't know its significance).

Douglas JRB-66A 52-2828. Seen is the prototype RB-66A in a vivid overall gloss red finish, with black radome and pylons, for its role as a GE CJ-805 engine test-bed, bailed to the General Electric Company. It was noticeable this year how many new GE engines were being developed, as they endeavored to catch up and pass Pratt & Whitney in the jet engine market.

(Top, Left) Douglas A3D-2 142243. This is the same A3D-2 as seen in 1958, but its paint scheme has changed a lot. Large areas of fluorescent orange have been added, plus stripes of the same color on the side of the large fuselage air brakes. Also, the large red number "243" formerly on the nose is now in the standard black color.

Martin B-57B 53-3864. The night-intruder version of the B-57 had a new nose housing the two-man crew in tandem, bombs carried in a rotating bay in the fuselage, plus extra racks under the wings, and either eight 0.5 inch guns or four 20 mm cannon in the wings—formidable armament, indeed! It is the 107th B-57B out of 202 built. However, it is still in the natural metal finish, with large areas of fluorescent orange applied, together with the ARDC tail marking.

(See also p. 65) Douglas DC-8-11, N8038D. Appearance of the commercial DC-8 at Edwards was an unusual sight, showing that initial flight testing of large jets was beginning to outrun manufacturers' plant capabilities. Seen in Douglas house colors, this is the fourth DC-8, having made its first flight on January 2, 1959. It eventually went to United Airlines as N8003U, racking up nearly 54,000 hours before being retired in April 1994.

Douglas C-133A, 54-135. Here is the very first Cargomaster, still at Edwards, in its overall natural metal finish, but now sporting acres of fluorescent orange paint!

Lockheed Jetstar, N329J. Seen in its second year at Edwards, the Jetstar is showing off a new white and fluorescent orange finish, but otherwise appeared to be the same.

(See also p. 68) McDonnell 119, N119M. A real surprise, this four jet utility transport was built to meet the same USAF requirement as the Lockheed Jetstar, but it was delayed a year while looking for a suitable engine. Originally designed for the Fairchild J83 jet, Westinghouse J34s were finally fitted, but Pratt & Whitney JT12s would have been fitted in any production versions. Unfortunately, the close proximity of the four podded engines to the ground caused the aircraft to be dubbed McDonnell's "vacuum cleaner"! Although otherwise satisfactory, this drawback caused the USAF to prefer the Jetstar, and the 119 ended up as a grounded type, despite McDonnell fitting it out as an executive jet, re-designated as the Model 220. Disgusted at the total lack of sales response, Mr. Mac ordered the sole prototype sold to an Arizona company, on condition that it would never be flown again or show any form of its McDonnell origin. The author saw it many years later, completely stripped of any finish or markings, at Deer Valley Field in Arizona.

NO
SMOKING

McDONNELL
NO
SMOKING

NO
SMOKING

Boeing NB-52A-BO-1, 52-003. This year saw the first appearance of a B-52 that was to become very famous as the launch carrier for the North American X-15 research aircraft. This was the fifth B-52 to be built, and the last of the three B-52As. It was natural metal finish all over, except for huge patches of fluorescent orange on the tail, nose, pods, and wing tips. A special cradle was fitted under the wing, inboard of the right hand inboard pods. The ARDC marking was carried on the vertical tail.

North American X-15, 56-6671. Here is the second of the three X-15s built, was to be seen at Edwards for the next three years while undergoing its incredible flight tests.

Helio L-28A 58-7025. First display at Edwards of the very distinctive high-wing light aircraft military version of the civilian STOL Helio H-395 Super Courier. Three were ordered in 1958, to be evaluated for operational techniques. This also included their suitability for supplying remote missile bases. More than 100 of various later versions of the type were used in the Vietnam war, as the U-10 (the change in the USAF and Navy aircraft designations is fully explained in Appendix 9).

Boeing KC-135A and B-52, (serials not visible). Boeing's biggest aircraft at this time are seen flying in a simulated in-flight refueling fly-past.

Vertol H-21, (serials not visible). Two of the H-21s assigned to ARDC are seen flying by the crowd, note their tail markings. None of the aircraft that flew by this year at Edwards were able to make the usual taxi-by, due to the tragic crash of a helicopter on the east end of the taxi-way. This involved a French-built Sud Djin single-rotor helicopter, which had a control system failure. Killing the pilot, this was the only crash to occur during any of the open house shows since the author's first visit in 1957.

1960

It was apparent that there were not so many aircraft on display this year, which was not surprising considering the cutbacks that the Eisenhower/Nixon administration had made in the services' budgets. The Korean conflict had been settled some seven years earlier, but the cold war with the Soviet Union was still very intensive. However, strange noises in the upper atmosphere from an orbiting Soviet satellite brought in a new era involving space, and this also diminished the funding available for new aircraft, to an extent not yet readily seen at Edwards. New aircraft displayed included the McDonnell F-101B, Lockheed F-104D, Douglas XF4D-1, Grumman F11F-1, North American A3J-1, Boeing B-52G, Convair B-58 (outside for the first time), De Havilland DHC-4, Lockheed NC-130B, Northrop N-156F, Cessna JT-37A, Hiller X-18 (in the hanger), and two North American GAM-77 missiles on the B-52G. The only new civil aircraft this year was the Convair 880. The aircraft displayed are listed below.

Manufacturer	Type	Popular Name	Aircraft Serial
Douglas	A3D-2	Skywarrior	142649
North American	A3J-1	Vigilante	146694
Boeing	B-47E-95BW	Stratojet	52-0514
Boeing	NB-52A-BO-1	Stratofortress	52-003
Boeing	B-52G-BW-105	Stratofortress	58-224
Convair	NB-58A-CF	Hustler	55-662
Douglas	RB-66A (CJ805)	Destroyer	52-2828
De Havilland	DHC-4	Caribou	57-3079
Douglas	C-54A-DO	Skymaster	41-37268
Lockheed	NC-130B-10LM	Hercules	58-712
McDonnell	F-101B-40MC	Voodoo	56-236
Convair	F-102A-CO	Delta Dart	54-1398
Lockheed	JF-104A-10-LO	Starfighter	56-748
Lockheed	F-104D-5-LO	Starfighter	57-1314
Convair	F-106A-CO	Delta Dagger	59-067
North American	F-107-1-NA		55-5120
Douglas	XF4D-1 (CJ805)	Skyray	587
Grumman	F11F-1	Tiger	138615
McDonnell	F4H-1 #11	Phantom II	145310
McDonnell	F4H-1	Phantom II	143389
Northrop	N-156F	Freedom Fighter	59-4988
Lockheed	T-33A-5-LO	Shooting Star	53-6045
Cessna	JT-37A	Tweet	56-3492
Northrop	YT-38	Talon	58-1192
North American	X-15		56-6670
North American	X-15		56-6671
Hiller	X-18		57-3078
Convair	880 #3		N803TW
Curtiss	Model D replica		NX5704N
North American	GAM-77	Hound Dog	59-2793
North American	GAM-77	Hound Dog	59-2795

(See alsoTop, Center p. 74) Boeing NB-52A-BO-1, 52-003, plus North American X-15, 56-6670, and North American X-15, 56-6671. This year, the NB-52A/X-15 combination was shown with the X-15 mounted on the carrier pylon. However, this was the first X-15, 56-6670, while the second one, 56-6671, was seen standing on its own landing gear, just off to the left side of the NB-52A. Both X-15s have the early long nose instrumentation probe, and are fitted with two of the original XLR11 rocket motors. Incidentally, during its third flight the second X-15 had made an emergency landing on Rosamond Dry Lake just over a year earlier, but shows no sign of it in the subsequent rebuild. Note the new red fluorescent finish, and markings on the right forward side of the NB-52A fuselage, indicating three captive flights and seven air-launches. Also seen is the huge drag-chute compartment, at the bottom of the rear fuselage, housing a 44 feet diameter FIST ribbon braking chute.

(See also p. 75, Top p. 76) Boeing B-52G-BW-105, 58-224. This was a real surprise, to see an operational B-52G on the ramp, together with two of the new North American GAM-77 Hound Dog air-to-ground missiles mounted, one under each wing. These missiles used a pylon fitted in the same position as that on the NB-52A. Both of the missiles were in a gloss white finish, while the aircraft was finished in the standard SAC natural metal finish, with gloss anti-nuclear flash white paint on all underneath surfaces.

80224
U.S. AIR FORCE

U.S.A
U.S. AIR FORCE

80224

Cessna JT-37A, 56-3492. First time seen at Edwards is the diminutive Cessna side-by-side basic trainer, gleaming in its polished metal and fluorescent orange finish. The "Tweet," as it became known, was the first U.S. trainer to have side-by-side seating since the Dayton Wright TW-3, circa 1922. Possibly, this was influenced by the contemporary British trainer aircraft of the Hunting Percival Provost/Jet Provost series. Probably the most remembered thing about the T-37 is the horrendous screech from its two Continental J69 turbojets (French designed Turbomeca Marbore engines), hence the nickname "Tweet" (after a real screechy bird in a current cartoon series).

Convair F-102A, 54-1398. The same GE engine test-bed F-102A as seen in the previous year. This time it was seen with the retractable J85 engine pod just slightly lowered, and the landing gear firmly planted on the ramp.

(Top, Left) Convair F-106A-CO, 59-067. This is the 67th aircraft from the last batch of 148 F-106As to be built, bringing total production to 277. The type was originally designated as the F-102B, in an effort to bring the original F-102A up to Air Defense Command's requirements for the main USAF all-weather interceptor. However, the changes were so extensive that it was finally re-designated as the F-106A. In any event, it was to prove to be the last interceptor specifically designed for the ADC mission, remaining in service until the end of the Cold War.

(See p. 78)

Convair NB-58A, 55-662. Production orders for 86 of the B-58 Hustler were finally placed in 1959, although the prototype had flown in November 1956. Development was protracted due to the radical design required for SAC's first supersonic bomber. This aircraft is the second one of the original eight test aircraft built, used by GE for testing the massive J93 in the under belly pod (also seen in the previous year). Thirty test aircraft were built, but only 86 service aircraft. The first operational unit, the 43rd Bomb Wing at Carswell AFB (right across the field from the Convair production plant), went into service in March 1960. Only two units were equipped with the B-58, and the type only lasted in service for ten years. It had a high accident rate, was extraordinarily expensive to build and operate, but did set a number of speed records during its service days. B-58s were always rather plain in their finish, being mostly natural metal. This one is seen with a bright red vertical tail, with the ARDC markings on the tail and forward fuselage. It was to serve at Edwards for many years, including as the main chase aircraft for the XB-70 program.

(See also p. 80) Convair 880, N803TW. We remarked in the 1959 review on the use of Edwards AFB for testing the new generation of large commercial jet transports. Convair also found it necessary to use Edwards, like Douglas with their DC-8, and here is the third of the 880 series, painted in General Dynamics' company colors, although registered as TWA N803TW. It first flew in March 1959, came to Edwards to carry out some of the more demanding FAA test requirements (such as heavy-weight short field landings), and was eventually delivered to TWA in October 1961. It flew for some 33,410 hours with TWA over a period of twelve years, later being in storage at Mohave Airport, California, for some eleven years before ending up at Atlantic City, NJ. There it was being used for burn tests by the FAA Fire Safety Branch. It was still there in 1996. Note the signs of extensive heating on the four GE CJ805 turbojet pods forward of the reverse thrust cascades. Incidentally, Boeing also ran their 707 through Edwards for similar reasons, but they never had any of their commercial aircraft on display at the annual Edwards open houses.

CONVAIR
880
N803TW
80712

N803TW

(Top, Left) Curtiss Model D Replica NX5704N. This accurate replica of a 1912 Curtiss Model D was built by Boeing engineer Peter M Bowers, and provides a vivid reminder of the enormous strides made in aviation when compared to the giant NB-52A/X-15 combination standing next to it. Bowers skillfully flew the replica twice during the display, despite the gusty cross-wind prevailing.

Bell HU-1A, 57-6099. Seen is the fifth of the first production version of the Army HU-1 Iroquois utility helicopter; little did we know that it would be produced in larger numbers than any other helicopter for the services, gaining ever-lasting fame for its role in Vietnam. Nicknamed the "Huey" from the original designation, it had been preceded by nine test X/YH-40s. The production version, seen here in Army test aircraft markings, had a one foot longer cabin to accommodate four stretchers, a wider crew door, and four inches greater ground clearance.

De Havilland DHC-4, 57-3079. De Havilland had gained wide acceptance of its L-20 Beaver in the U.S. Army, and its new twin-engine design was selected as the AC-1 Caribou in 1957. Here is the first of five aircraft for evaluation, seen being towed into position. Painted gloss olive drab all over, with white markings on the nose and yellow-orange ones on the vertical tail, the type had originally been registered as CF-LKF-X in Canada before being delivered to the Army. It proved to be the only new piston-engine aircraft on display, making its first flight in March 1959. 159 were bought by the Army, eventually ending up with a new designation of CV-2 in 1962. They were all turned over to the USAF at the beginning of 1967, being again re-designated as C-7s. They became extremely popular in the Vietnam war for their versatility.

Douglas C-54A-DO, 41-37268. The C-54A continued to be used at Edwards as a transport, and the old perennial showed up once more, but this time with its designation changed to the normal one of C-54. However, it did show lots of fluorescent orange added to the wing tips, nose, and rear fuselage.

Douglas RB-66A, (CJ805), 52-2828. This year, the venerable RB-66 was carrying the new GE CJ-805 aft fan engines under the wings. These had long, large diameter pods, which looked like they might present a problem in a cross-wind landing.

Douglas A3D-2, 142649. A later production version of the "Whale" was present this year, the 125th one from the -2 line, from the second largest batch of them to be built. It had lots of fluorescent orange on it and sported a new radar tail fairing. Note the fine color stripes on the engine pods and the aft fuselage air brakes.

Douglas XF4D-1 (CJ805), 587. Among the large number of engine test-beds was the second prototype XF4D-1, BuNo. 124587. Initially fitted with an Allison J35 engine as the intended XJ40 engine was not available, this aircraft became famous when finally fitted with the reheat XJ40-WE-8. In that form it took the world speed record away from the Supermarine Swift Mk.4 (for which the author had designed a major control system), on October 3, 1953, and only two weeks later it also broke the 100-kilometer closed course speed record at Edwards. It was distinctive in being the only F4D fitted with a sliding canopy (rather than the usual clam-shell type). Finished in a highly polished natural metal finish, with bright red trim, it was fitted with a scale model of the aft-fan CJ805-23 engine pod mounted under the right wing. This pod was intended for use on a new version of the Sud Caravelle, but it did not enjoy market success.

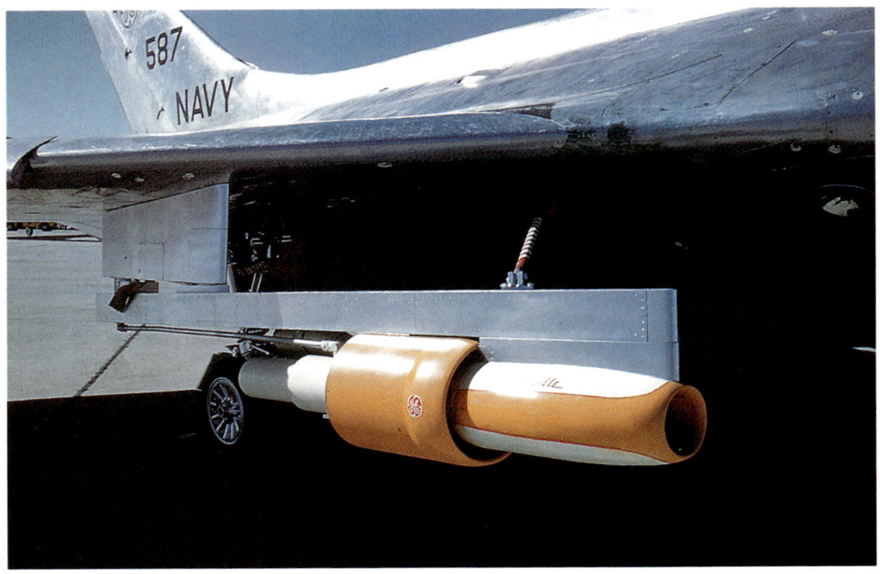

Grumman F11F-1, 138615. A stock production version of the Tiger is seen with lots of fluorescent red on wings, nose, and tail. A rather smug-looking tiger was painted on both sides of the vertical tail. The Tiger became famous primarily through its use by the Navy Blue Angel acrobatic team, but in fact was being phased out of front-line service during 1959, being sent to the Navy Advanced Training Command. Its major drawback was the lack of fuel, giving a rather short range for a carrier based fighter.

(See also p. 86) Lockheed NC-130B-10LM (BLC), 58-712. This was a very interesting STOL test version of the C-130, fitted with blown flaps having a full 90 degrees of travel, together with drooping ailerons. The two pod mounted Allison YJ56 gas generators were mounted on the pylons normally used by the long-range fuel tanks, and provided high speed air which was blown over the flaps, ailerons, rudder, and elevators to get the boundary layer control required for very low speed flight. It is in a highly polished natural metal finish, and carries liberal amounts of fluorescent orange in addition to the Arctic-red paint on the vertical tail, which had a rudder with a forty percent increased chord. Not usually seen in public was the massive amount of wool-tufts mounted all over the aircraft, to visually discern the airflow changes caused by the high speed airflow. Intended as the prototype for a STOL version of the C-130, it made its first flight in February 1960 and was eventually handed over to NASA for use in the Earth Survey program, registered as N929NA.

U.S. AIR

80712

Lockheed T-33A-5LO, 53-6045. In contrast to the NC-130B, is this completely stock version of the T-33 trainer, part of the ARDC fleet at Edwards (note the markings on the air intake, and the tip tanks in fluorescent orange, with black trim).

Lockheed JF-104A-10-LO, 56-748. First of the -10 block version of the F-104A, this is one of a batch allocated to ARDC at Edwards to sort out continuing problems with early versions of the F-104. Some of these included unreliability of the 20mm M-61 Vulcan gun, air intake flow, stability in certain attitudes, etc. Like the F-101, the F-104 suffered from pitch-up problems, caused by the T-tail. However, probably its worse feature was the downward ejecting seat: this feature killed the USAF's best test pilot, Iven Kincheloe, when he had an engine flame-out at a low altitude just after takeoff. All of these problems were eventually solved (except the inherent pitch-up) on later service versions of the aircraft. This particular F-104A has had a large ventral fin fitted; this became standard on all versions of the aircraft. These problems caused the F-104 to have a very lengthy test program at Edwards, and many aircraft were lost. As a result, USAF finally only bought 296 of all versions, rather than the originally planned 722. To try and recover their large investment in the type, USAF actively supported Lockheed efforts to sell the aircraft overseas, mainly in Europe. One unfortunate result of this effort was to result in a delay in getting the new Northrop F-5 into production and service overseas, despite it being a better aircraft for most countries than the highly sophisticated F-104.

Lockheed F-104D-5, 57-1314. Here is the final two-seat trainer version of the USAF F-104 family, with two tandem seats, and an extended-chord vertical fin and rudder protruding aft of the tailpipe. Unlike the other F-104 on display, this one was painted with lots of fluorescent orange; note how it is beginning to wear off due to the high performance of the aircraft.

McDonnell F-101B-40MC, 56-236. We get our first sight at these displays of the two-seat all-weather fighter version of the F-101. It had been ordered as a stop-gap aircraft, to fill the need for a truly supersonic interceptor before the highly redesigned F-106 would be available (the F-102A then in service was barely supersonic). However, it was finally delivered to units some 18 months later than planned, due to change of engines from the planned Wright J67 (Bristol Olympus built under license) to P & W J57s. It is the fifth F-101B built (the block numbers for the 101s started with the first production version of the F-101A, continued through all versions built), and is in natural metal finish with a small amount of fluorescent red on the tail unit, and in a fancy scalloped pattern on the white nose-radome (all fluorescent finishes were applied over a base coat of white paint).

Northrop N-156F, 59-4988. Although the T-38 ended up being built first, Northrop strongly pursued the N-156F fighter version, and was soon rewarded by a contract for two prototypes from the Department of Defense. This is the second prototype, seen next to the T-38. Finished in an overall gray, with the usual fluorescent orange bands, it displayed the name for the type on the nose, "Freedom Fighter." (The author was heavily involved with testing of both the N-156F and the T-38 at this time. He was responsible for the drag chute system on the N-156F and the spin recovery chute system on the T-38: more details of this will be found in Appendices 3 and 4. Both proved to be rather exciting efforts, causing him to spend long, rewarding hours at Edwards).

McDonnell F4H-1, 145310, #11. As indicated by the small number 11 on the vertical tail, the statically displayed F4H-1 was the eleventh to be built. Externally similar to the #1 aircraft seen in the previous year, the 23 test aircraft were used to evaluate several changes. These included a new canopy with better vision from the rear cockpit, larger radome for more powerful radar, modified intakes with new movable forward ramps and fixed aft ramps with bleed air holes, and the installation of boundary-layer air blowing over the leading and trailing edge flaps. This aircraft is seen carrying three massive long-range fuel tanks under the belly and inboard of the wing-fold. Their vivid fluorescent red finish contrasts vividly with the light gull gray and white finish of the F4H-1. In the taxi-by photo we see the fourth F4H-1, BuNo. 143389, painted similarly to all of the test batch. It had just given a dramatic demonstration of its maneuverability and high roll rate.

(See alsoTop, Center p. 91) North American A3J-1, 146694. First showing of the very sleek new North American was of the third aircraft, also the first production model. The prototype had first flown in August 1958, and delivery of production aircraft to VAH-7 had begun in 1960. Configuration of the Mach 2.1 design included a very thin in side-view fuselage, parallel for most of its length, but wide in plan-view aft of the two intakes, and a high-mounted broad-chord swept wing. The control system was very advanced, using an all-moving vertical tail and differential moveable horizontal stabilizers to give both pitch and roll control. There were no ailerons, and blown flaps were used. Variable geometry intakes were provided for the two GE J79 engines. The Vigilante had a very unusual bomb bay between the engines, with a nuclear weapon being ejected rearward. However, the system never performed satisfactorily, and resulted in the type being converted to a reconnaissance role, as the A3J-3P (later RA-5C). There have been many rumors that the A3J configuration was copied by Mikoyan for the Mig 25, but as that aircraft was designed for Mach 3 to combat the new Lockheed A-12 fighter (more of which in the 1965 chapter), there is only a superficial resemblance. The use of twin vertical tails on the MiG 25 (unlike the single one on the A3J), led to extensive research in the U.S., resulting in the adoption of twin verticals on the new F-14, F-15, and YF-17/F-18 fighters (the author was involved in this to some extent at the time). (The A-11/YF-12/SR-71 family also had twin vertical tails, but they were required more for stability and control at very high Mach numbers, like the XB-70, rather than for fighter aircraft maneuvering control at high angles of attack).

North American F-107-1-NA, 55-5120. The XF-107 was to be seen for the last time at these displays this year, and outwardly only differed from last year's version in removal of "NASA" from the yellow band across the vertical tail, presumably showing that NASA had completed its test program and handed the aircraft back to the USAF.

Northrop YT-38, 58-1192. Seen outside for the first time, this is the second prototype YT-38, looking no different than the first one seen in the previous year. Flight tests had gone extremely well for the type, and it was now in full production for the USAF. Three T-38As put on a spirited fly-past this year, finished in the standard white overall with fluorescent trim finish. Derivation of this first USAF supersonic trainer was quite involved, coming from a privately-funded light fighter design, the N-156. When North American was awarded the XF-108 contract, instead of Northrop, it was rumored that the T-38 contract was given to Northrop as a consolation (as Northrop had tendered a very advanced design for the Mach 3 all-weather fighter program). Whatever the truth, Northrop ended up by being the big winner, building no less than 1,189T-38s, while the XF-108 was canceled prior to being built.

Ramp display view. For the first time, the author was able to photograph the ramp display at Edwards from the roof of the building on the west side of the huge hangar. It is apparent how large this ramp is from this view looking west over only a short portion. The clear visibility in the Antelope Valley is also evident from the mountains visible to the south; unfortunately, the visibility deteriorated considerably over the next 25 years or so.

1961

The lack of new aircraft and the total of aircraft displayed at Edwards continued to shrink, and this year's showing was very thin. However, I was fortunate enough to be taken down by the main runway and photographed the flying display aircraft, both landing and taking off. This led to some interesting photographs, especially as there was a very notorious type on display for the first time. New aircraft were very scarce, but they included the Douglas A4D-2N, Republic F-105D, North American T-39, plus the notorious Lockheed U-2, and the Cessna U-3. It was also noticeable that there was not quite as much fluorescent paint being applied, as it was costly to apply and maintain, and had apparently not had the desired results in making aircraft more visible. The aircraft that were present are listed below.

Manufacturer	Type	Popular Name	Aircraft Serial
North American	A3J-1	Vigilante	146701
Douglas	A4D-2N	Skyhawk	145069
Boeing	B-47E	Stratojet	52-221
Boeing	B-47E-95BW	Stratojet	52-0514
Boeing	NB-52A-BO-1	Stratofortress	52-003
Martin	B-57E	Canberra	55-4258
Martin	JB-57E	Canberra	55-4267
Convair	NB-58A	Hustler	55-662
Douglas	RB-66B	Destroyer	52-2828
Fairchild	C-119F-FA	Flying Boxcar	51-8041
Lockheed	C-130A-1-LM	Hercules	54-1624
Lockheed	C-130A	Hercules	54-1626
Lockheed	C-130A	Hercules	53-3135
North American	F-86F-25-NH	Saber	52-5412
North American	F-100F-15-NA	Super Saber	56-3963
McDonnell	F-101B-120MC	Voodoo	59-469
Convair	TF-102A	Delta Dart	54-1354
Lockheed	F-104A	Starfighter	56-0760
Lockheed	F-104A	Starfighter	56-0764
Lockheed	F-104A	Starfighter	56-0766
Lockheed	JF-104B	Starfighter	56-3721
Republic	F-105D-5RE	Thunderchief	58-1150
Convair	F-106A-CO	Delta Dagger	56-467
Douglas	XF4D-1	Skyray	587
McDonnell	F4H-1F	Phantom II	145310
Sikorsky	H-37A	Mojave	55-621
Lockheed	T-33A-5-LO	Shooting Star	52-9796
Cessna	JT-37A	Tweet	56-3492
Northrop	T-38A-NO	Talon	58-1192
Northrop	T-38A-NO	Talon	59-1597
North American	T39A-1-NA	Saberliner	59-2872
Lockheed	U-2D		56-6954
Lockheed	U-2A		56-6701
Cessna	U-3A	Blue Canoe	57-5849
North American	X-15		56-6671

(Top, Left) Sikorsky H-37A 55-621. We last saw the H-37A in 1958, and this is one of the original forty ordered in 1955. After the Army had completed taking delivery of their total of 94, they began a program to modernize 90 of them. This involved installing redesigned cabin door and cargo hatches, crash-resistant fuel tanks, plus other changes. The resulting upgraded helicopters were re-designated as H-37Bs, the first one being delivered in June 1961.

Cessna JT-37A 56-3492. The same aircraft as shown in 1960, the only change appears to be an even brighter polish of its natural metal finish!

Cessna U-3A 57-5849. Definitely new this year was the Cessna U-3A, a military version of the very popular Cessna 310. It was originally selected as the L-27A, with 160 being purchased. They were popularly known as "Blue Canoes," probably due to their stock color scheme. The tip tanks were painted in the brilliant fluorescent red finish with the ARDC Edwards black lightning bolt along them.

Lockheed T-33A-5-LO 52-9796. Another year, another stock ARDC T-33A on display, but with last year's fluorescent orange replaced with fresh fluorescent red.

Lockheed JF-104B 56-3721. We had not seen a F-104B on display since 1958, and this one was very plain, without any fancy fluorescent colors. Assigned to ARDC, it now showed the standard white paint on the wing surfaces and the tip tanks.

Lockheed C-130A-1-LM 54-1624. Seen is the fourth C-130 from the 1954 batch of twenty aircraft, the first real production order for the aircraft. It is immaculate in natural polished metal, with Arctic red paint on the vertical tail, and fluorescent red bands on the fuselage nose and tail, and on the wing tips. Like many of the early C-130s, it does not have the later nose radar fitted, so it retains the original bluff nose contours.

(See also p. 98) Lockheed U-2D 56-6954. The infamous U-2 had burst into public view when Gary Powers was shot down by a Soviet SAM on a mission over Russia. Many people knew about the aircraft before then, but kept quiet about it, as the cold war was getting pretty hot at the time. There was no point in hiding the aircraft any longer, but we were still very pleased to see it at Edwards that year. The U-2D on static display had a very unusual finish of flat black over the forward portion of the aircraft and above and below the wings; the lower forward portion of the aircraft was in light gray. The aft fuselage and tail unit were finished in natural metal. In contrast to the operational U-2s, this one had huge letters spelling out U.S. AIRFORCE on the aft fuselage (note the lack of space in the spelling). White wing walkway markings were augmented by a horde of red "NO STEP" markings, while the rest of the airframe was adorned with a mass of other stenciled markings. There was an interesting "Smokey Joe" cowboy drawing on the vertical tail, showing that it was assigned to the Edwards-based 6512th Test Group of the Air Force Flight Test Center. This is a two-seat version, used for many test purposes. It is believed that in the late 1960s it eventually went to the 100th SRW, as a black painted U-2C version. The mass of stenciled cautionary warnings were necessary due to the U-2's extremely light construction, as it was really a very large jet-powered glider, mated to a F-104 cockpit. An extremely clever design for its special purpose. We were to see another version of the U-2 fly that day, more on that later.

U.S.AIRFOR

(See also Top p. 100) North American T-39A-1-NA, 59-2872. First time seen at Edwards, the Sabreliner came from a requirement from the USAF requiring the manufacturers to design and fly a private venture utility trainer experimental (UTX). This was issued in August 1956, and North American flew a suitable prototype in September 1958. An initial production order for seven aircraft was awarded in January 1959, after flight test evaluation at Edwards. This is the fifth aircraft; the author flew the seventh one on May 11, 1961, and found it to be a very good performer (See Appendix 7). 149 were eventually built, being used by Air Training Command, SAC, Systems Command, and USAF Headquarters. Six T-39Bs were used as trainers for the radar system used in the F-105, while the T-39D and T-39E went to the Navy. The T-39 configuration was unusual in that it retained a low-set horizontal stabilizer with the aft mounted engines; most other aft-engine types of the era used the high T-tail of German origin.

(Left, Bottom) Convair F-106A-CO, 56-467. When this F-106A was initially spotted on the ramp, it seemed to be only another one of the production aircraft, finished in aircraft gray overall. However, before the crowds appeared a notice was put on the right side of the fuselage. This stated that the F-106A had broken the World speed record on December 15, 1959, at a speed of 1,525.9 mph over a course at Edwards, flown by Major Joseph W. Rogers. Why the actual record-breaking aircraft was not exhibited at the 1960, nor the 1961, display is a mystery, but it brought attention to the number of such aircraft being tested at Edwards AFB.

(Left, Center) Convair NB-58A, 55-662. The NB-58A appeared for the second year, but this time it did not carry the huge J93 engine test pod underneath the belly. It was also no longer bailed to GE, but was back in the hands of ARDC. The only concession to color was the overall red vertical tail, with a very large black ARDC flash on it.

Douglas XF4D-1, 587. Present once again, still bailed to GE, but this year with only a large drop tank fitted under the left wing. Nevertheless, the comparison between the speed record holder of 1956 and the current F-106A was instructive, showing the very rapid pace of development in those years.

(Top, Right) Republic F-105D-5-RE, 58-1150. We had seen the earlier F-105s previously, but this was the first sight of the definitive single-seat version. The new nose radome, with a long nose probe, changed the appearance of the front-end considerably. This was the fifth one to be built, and was basically in natural metal finish, with some added fluorescent patches. Unusual was the long-standing Republic use of dark olive drab in front and aft of the canopy (with a large cream dielectric panel aft of the canopy). Some months prior to this display, there had been a special demonstration of the available USAF fighter aircraft, to a visiting delegation from the NATO air forces. The author was working at Edwards that day, and will never forget the sight and sound of the F-105D roaring around in full afterburner power at about 100 foot altitude, making very tight turns. Most impressive!

Douglas RB-66A (CJ805 aft fan), 52-2828. This appeared to be exactly the same as in the previous year, at least in all external respects.

(Right, Center) Douglas A4D-2N, 145069. Another first this year was the new A4D-2N, adding new all-weather and terrain-clearance radar to the "Scooter." This changed the nose contours, and lengthened it by 9 inches. It is carrying three of the large Bullpup AGMs, and is finished in both fluorescent red and orange in different areas. This turned out to be the most produced version of the aircraft, no less than 638 being delivered to the Navy, the first becoming operational with VMA-225 in February 1960.

Boeing B-47E, 52-221. Assigned to the AFFTC, this B-47E had the white anti-flash finish under the fuselage only, and had a small amount of fluorescent orange paint on the fuselage and wing-tips. The new FTC insignia were on the nose and the large vertical tail flash.

Boeing NB-52A-BO-1, 52-003. The perennial test launch aircraft had a lot more captive and flight launch markings on the nose, plus a rather elegant insignia and caption for the "High and Mighty One." Very appropriate for this particular aircraft.

Martin JB-57E, 55-4267. This B-57E (not displayed since 1957) now showed the usual fluorescent red patches. However, careful examination of it shows that the bomb-bay has black and yellow stripes running fore and aft, possibly causing the "J" prefix to its designation.

Fairchild C-119F-FA, 51-8041. It was surprising to see this twin-boom transport at this late date, but it has a new feature in the form of a massive beaver-tail fairing to the fuselage. First flown on C-119G, 53-3213, this tail was retroactively fitted to 68 C-119Fs and C-119Gs, causing them to be re-designated as C-119Js. Note how the engine exhaust has dirtied the fluorescent orange paint on the rear end of the booms.

McDonnell F-101B-120MC, 59-0469. With a very plain natural metal finish, and a total lack of unit markings, this F-101B looks like it had just rolled off the McDonnell production line.

(Top, Right) McDonnell F4H-1F, 145310. This year, the F4H-1F was shown with a full complement of six AIM-7 Sparrow air-to-air missiles fitted semi-recessed in the fuselage belly, and under the inboard wing pylons.

Northrop T-38A-NO, 59-1597. This is a completely stock T-38A, with the newly adopted Training Command white finish overall, plus the usual fluorescent red bands. Note the very extensive stenciling on the forward fuselage.

North American F-86F-25-NH, 52-5412. Last displayed in 1959, the F-86F was getting rarer at the open houses as they were superseded by later types. However, this one is as immaculate as ever, with fresh fluorescent red, but still carrying the old ARDC insignia.

North American X-15, 56-6671. Once again, the second X-15 appeared, but this time it was fitted with the massive XLR99 rocket motor. After North American contractor flights had been completed in December 1960, NASA/USAF/U.S. NAVY research flights began, with Major Robert White at the controls. Prior to this Edwards display, he had already taken the X-15 to a speed of 2,905 mph, and an altitude of 169,600 feet. Graphic evidence of this performance can be clearly seen in the scorching of the upper vertical tail, and its markings. Also note the cover on the nose of the aircraft; this gave ground protection to the remarkable Northrop "Q-ball," which provided angle-of-attack, sideslip, and dynamic pressure information, despite the extreme heat and pressure it suffered.

North American A3J-1, 146701. This, the eighth production A3J-1 (the number "10" on the vertical tail includes the two prototypes), also showed signs of lots of heat around the rear fuselage tail fairing. Although lots of testing had been done on this unique rear-ejecting bomb-bay, it never worked as specified, and was finally abandoned when the Navy strategic bombing role was deleted by the Department of Defense. The fluorescent orange paint also looks the worse for wear and has been supplemented with fluorescent red.

This year, the author had been fortunate enough to get permission from the Air Force to go out to the main runway at Edwards, and photograph all of the aircraft taking off and landing for the flight portion of the open house. It was very hot (and noisy), but he was lucky enough to see, and photograph, the first public flying display of the U-2. The resulting photographs were the first ever published showing the unique pogo-stick landing gear of the U-2 in use. Incidentally, he was told that he could photograph any aircraft using the runway, except for one type. If that was to appear, he would get orders to put his cameras down on the ground and walk away ten feet from them, until given permission to use them again. The aircraft? He never found out!

Lockheed U-2A, 56-6701. Still to be seen at the SAC Museum at Offutt AFB, Omaha, Nebraska, this aircraft was assigned to AF Systems Command after its initial use by the CIA. It is seen in natural metal finish, with large USAF lettering, and fluorescent red. Note the two different insignia on the nose and tail flash.

Lockheed C-130A, 53-3135 and 54-1626. Two of the early bluff-nosed C-130As took part in the flying, wearing different paint work. The earlier one shows very distinct signs of wear on the nose fluorescent orange paint, while the second one does not have the Arctic red vertical tail paint.

Northrop T-38A-NO, 58-1192. This is the second T-38A prototype, still having the rear cockpit painted over to protect the test instrumentation package inside.

Boeing B-47E-95BW, 52-0514. Seen in the static display in 1960, this B-47E gets away with a characteristic trail of black smoke from its six GE J47 jets. It took a fairly long run to get off the runway.

Boeing NB-52A, 52-003. At last, we got to see a B-52 close-up during takeoff. Lots of smoke and noise marked its departure, but the surprise was the very steep climb-out, with the wings flexing upwards under the air loads. Of course, it was not carrying any payload, nor much fuel.

(Top, Left) Martin B-57E, 55-4258. The B-57E was the first aircraft ever built new as a target tug for the USAF, sixty-eight of them being delivered from August 1956. However, they could be converted back to bomber configuration fairly quickly if required. The tractable flying qualities of the B-57 family can be seen in its short take-off and landing runs, evident here.

(See also Top p. 113) Convair NB-58A, 55-0662. The lightly laden B-58 also got away very smartly, accompanied with a great deal of noise from its four GE afterburning J79s. It later used its large drag chute to reduce its considerable landing roll.

Lockheed F-104A, 56-0764, 55-?, 56-0760. Several F-104As flew and demonstrated their marked acceleration, and maneuvering capabilities, also making a remarkable amount of noise from their afterburning GE J79s.

(Top, Right) North American F-100F-15-NA, 56-3963. The last production version of the first Century series fighters, the F-100F two-seater, did not look so hot after seeing the later Century series fighters perform, its longer takeoff run being noticeable. Its afterburning P & W J57 still made a lot of noise!

Convair TF-102A, 54-1354. This trainer version of the third Century series fighter got away spritely, due to its large area delta wing; however, it was only subsonic, mainly due to the drag of its large side-by-side cockpit.

1962

There turned out to be a few more aircraft displayed at Edwards this year, showing lessened use of the fluorescent red and orange paints. Actually, it now seemed to be used mostly for trim on aircraft primarily based at Edwards, as the newer combat aircraft shown were in plain natural-metal finish. There were few new aircraft displayed, these included the Hiller H-23, Sikorsky H-37B, De Havilland L-20, North American T-28A and T-28B, Cessna T-37B, Bell HU-1, North American X-15 No. 3, and two privately owned replicas, a Ryan N-Y-P and a Eberhardt S.E.5. As can be seen, these were mostly aircraft being used or being developed for close support of the Army in minor combat actions, as the USA was now sending "advisors" to SE Asia to try and quell local Communist uprisings. The aircraft displayed are shown in the following listing.

Manufacturer	Type	Popular Name	Aircraft Serial
Boeing	TB-47B-26BW	Stratojet	51-2075
Boeing	B-47E-95BW	Stratojet	52-0514
Boeing	NE-52A-BO-1	Stratofortress	52-003
Boeing	NB-52B-BO	Stratofortress	52-008
Boeing	C-135B-BN	Stratotanker	61-0331
Martin	B-57E-MA	Canberra	55-4235
Martin	B-57E-MA	Canberra	55-4262
Convair	B-58A-10-CF	Hustler	59-2456
Douglas	C-54A-DO	Skymaster	42-72557
Lockheed	C-130A-1-LM	Hercules	53-3135
Lockheed	C-130A-1-LM	Hercules	54-1624
North American	F-100D-1NA	Super Saber	54-2127
North American	F-100F-1NA	Super Saber	56-3726
Lockheed	F-104A-10-LO	Starfighter	56-0754
Lockheed	F-104A-10-LO	Starfighter	56-0760
Lockheed	F-104A-10-LO	Starfighter	56-0762
Lockheed	F-104A-15-LO	Starfighter	56-0766
Lockheed	F-104B-5-LO	Starfighter	57-1315
Republic	F-105D-5RE	Thunderchief	59-1718
McDonnell	F4H-1	Phantom II	145310
Vertol	H-21B	Workhorse	51-15854
Hiller	H-23D-UH	Raven	55-4062
Sikorsky	H-37B-SI	Mojave	54-998
De Havilland	L-20A-DH	Beaver	58-2011
Northrop	N-156F	Freedom Fighter	59-4987
North American	T-28A-NT	Trojan	52-1234
Convair	T-29B-CO		51-7916
Lockheed	T-33A-5-LO	Shooting Star	53-5101
Cessna	T-37A-CE	Tweet	57-2331
Northrop	T-38A-NO	Talon	59-1595
Northrop	T-38A-NO	Talon	59-1598
North American	T-39A-1-NA	Saberliner	59-2869
Bell	HU-1-BF	Iroquois	57-6100
Lockheed	U-2A-LO		56-6701
Lockheed	U-2D-LO		56-6954
Cessna	U-3A-CE	Blue Canoe	57-5849
North American	X-15		56-6672
Ryan	N-Y-P replica		N-X-211
Eberhardt	S.E.5 replica		A.S.22-296

Bell HU-1A, 57-6100. This is the sixth Iroquois to be delivered, deliveries of 173 of this version being completed in March 1961. This is very much a test version at Edwards, resplendent in fluorescent red trim, in addition to the standard Army yellow bands. It is carrying three small inert (blue color) missiles on the right pylon. The type was among the first to be deployed to Vietnam, by then carrying two 0.30 in machine guns and sixteen 2.75 in air-to-ground rockets.

BoeingTB-47B-26BW, 51-2075. We had last seen this particular B-47E in 1959 (as a JB-47B), but it was now designated as a TB-47B, a training version. It is seen totally devoid of any markings or additional paint of any kind, looking extremely bare! Possibly, it was being prepared for delivery to another unit.

Boeing B-47E-95BW, 52-0514. In contrast, the flying display B-47E carried fluorescent red bands on the nose, rear fuselage, and at the wing-tips. It also showed the ARDC fin flash and the Edwards FlightTest Center markings on the nose.

(Right, Center) Boeing NB-52A-BO-1, 52-003. Once again, for the fourth year in a row, the NB-52A was shown, this time with the third X-15, 56-6672, mounted on the pylon. Only visible change in the NB-52A were a few extra flight markings on the right side of the fuselage. However, this was the first time that the third X-15 had been seen, fitted with the large XLR99 rocket motor. This had been rebuilt after a disastrous on-ground explosion in June 1960, and had made its first flight as recently as December 20, 1961. It had subsequently made three more flights prior to this open house.

Boeing C-135B, 61-0331, MATS. (4932). Here we see the first of thirty cargo/passenger versions of the KC-135 family built for the Military Air Transport Service. In polished natural metal finish, with fluorescent red bands on fuselage and wing-tips. The standard blue MATS band appeared across the later, taller, vertical tail, and the name was spelled out along the top of the fuselage. Large black panels with white crosses were marked on the upper fuselage sides for photo-interpretation during performance tests. The C-135Bs were fitted with the later P & W TF33 turbofan engines, giving much improved performance. Not the least of this was the deletion of the heavy water-injection (smoke-generating) system used at takeoff by the J57 powered versions.

Cessna U-3A, 57-5849. Last years new "Blue Canoe" appeared again, but had been shorn of the attractive blue paint scheme that led to its nickname (which lasted long after the paint was removed).

CessnaT-37A, 57-2331. Looking exactly as it had in previous years, the now standard basic trainer was being used in a new style. In April 1961, the USAF had changed to all-jet training. This had resulted in high costs, due to training pilots who failed the primary flight phase. Eventually, the training program went back to using a primary phase on piston-engine CessnaT-41As, in 1965.

Convair T-29B, 51-7916. It was surprising to see this USAF navigation/radar trainer at Edwards, long after it had entered service in 1952. Some 105 of this version were built, carrying ten student navigators and four student radar operators. Smartly finished with a white top, and black cheat line, it had fading fluorescent orange bands in the usual places.

(Left, Bottom) Convair B-58A-15-CF, 59-92456. This year, we were treated to a full-blown production version of the Hustler, complete with its two-part external weapon pod. This was shown on a wheeled dolly, under the belly. Unlike the majority of service B-58s, which normally carried the usual SAC band around the fuselage, this one has no distinguishing markings or paint, being natural metal finish overall.

De Havilland L-20A, 58-2011. This late-production Beaver carries full Army colors and markings. Some 968 L-20As had been delivered by the end of 1960, their purchase from Canada requiring special Congressional approval. During the Korean War, it had been nicknamed "the General's jeep," due to its extensive use by Generals Van Fleet, Mark Clark, and Ridgeway. Finished in gloss olive drab, this example carries large areas of fluorescent red.

(Left, Bottom) Sikorsky H-37B, 54-998. First seen in 1958 as an H-37A, this example has been modernized and brought up-to-date as a H-37B. Its gloss olive drab Army finish is in better shape than that seen on previous examples.

Eberhart S.E.5E, A.S. 22-296. The British designed S.E.5 fighter was supplied to the Air Service, and was going to be produced by Curtiss. They assembled 57 modified British-built airframes, near the end of WWI. The Eberhart Steel Products Company rebuilt 50 of them as advanced trainers in 1922 and 1923 for the Air Service. Though this aircraft appears to be true to original, the color scheme is not quite correct. It should have the yellow finish on only the upper surfaces of the wings and horizontal tail, and on the fin. The national insignia should also appear on the lower surfaces of the lower wings. Despite these minor inaccuracies, the type was a very welcome sight at Edwards, showing just how far design had evolved in the intervening forty years.

(Top, Right) Vertol H-21B, 51-15854. Some 163 H-21Bs were built for the USAF, and here is the very first one, some ten years after its first flight. It is carrying the optional extra fuel tank under the belly. The flight shot shows another of the type flying over the severely burnt remains of no-less than the world speed record holding F-106A! This had been used for the usual fire-fighting display this year.

Douglas C-54A-DO, 42-72557. Once again a C-54A showed up, but this one was a different one than those seen in previous years. It was beautifully finished, with fresh fluorescent red paint in the usual areas. However, unlike previous examples, it carried the name "Edwards Air Force Base" on the nose wheel doors. It is also fitted with radar in the nose.

Lockheed U-2A, 56-6701. This aircraft took part in the flying display again. There were no visible differences in its appearance from its debut last year, and it gave the same exciting flight display. Note how the auxiliary support pogos under the wings are trailing in the aft position as the aircraft is being towed out for its flight, but are in the opposite position when it is taxiing back under its own power.

(Top, Right) Lockheed U-2D, 56-6954. Again on static display, the U-2D displayed some minor, but significant, changes to its markings. The "Smokey Joe" Edwards unit marking had been removed from the vertical tail, although the ARDC fin flash remained. Incidentally, the display of the aircraft serial number on the tail is incorrect: strictly speaking, it should appear as "56954," not as painted, "566954." It has been noticed that the serial number presentation on U-2s differed from the regulations many times: in fact, many appeared to have changeable digits, probably to fit security needs. Closer examination of the airframe revealed two new di-electric panels had been fitted in the belly, just forward of the front landing gear doors.

Ryan N-Y-P replica, N-X-211. This had been built for the famous movie covering Lindberg's history-making 1927 flight to Paris. Once again, it helped to show the tremendous progress made during the previous thirty-five years.

Lockheed F-104A 56-754, 56-0762, 56-0760, 56-0766. Most of this early batch of F-104As had been seen at Edwards during the last five years, and showed little or no external changes made during the type's protracted development. At this stage, they were being used primarily by the Test Pilots School, and as chase aircraft during testing of other high performance aircraft. The close-up shot of the left engine intake shows the new Air Force Systems Command insignia being applied to all of the former ARDC aircraft. Note that the intake shock cone was a very fixed item, rather than being moveable as on nearly all other Mach 2 plus aircraft. This was one of the chief areas requiring a great deal of development on the aircraft, causing a great many ECPs (engineering change proposals) to be written. Note the different positioning of the serial number on the tail of some of the aircraft.

Lockheed F-104B-5-LO, 57-1315. This is the two-seat version of the F-104A, used to transition pilots into the type. Most of the Century series fighters needed two-seat trainer versions (however, note that the F-105F two-seat version was used to train pilots already skilled in the F-105D how to handle later, very complex, weapons systems). Note the red covers placed on the leading edge of the wing to prevent injury to ground crew members.

Republic F-105D-5-RE, 59-1718. The F-105 was one of the first aircraft to carry the new TAC (Tactical Air Command) badge and flash on the vertical tail. Otherwise it carries no distinguishing marks, nor does it have any fluorescent red patches applied. The Air Force was already beginning to find that fluorescent paint was not a very effective anti-collision system, and its use cost a lot of time and money. Quick fading of the fluorescent color was a particularly bad feature. Thus, its use was beginning to be confined to certain aircraft operating at Edwards AFB. Production of this version of the F-105 was held up early in 1962 by a strike at Republic, and in June 1962, all F-105B and D aircraft were grounded after two major accidents at Nellis AFB. The corrective program, called Look Alike, took much longer than expected, not ending until mid-1964 (but the aircraft were back to flying status long before then).

Lockheed T-33A-5-LO, 53-5101. A different airframe from the T-33s seen during the last four years, this one was just as highly polished, and also carried some very fresh fluorescent red on the tip tanks. However, it also carried no unit markings of any kind; the USAF was beginning to go through a phase of rendering its aircraft almost totally anonymous, possibly due to the "Cold War" getting very warm at this time.

(Top, Left) Martin B-57E, 55-4262, 55-4235. Two more of the prolific B-57Es were seen at Edwards this year. The static display one was in natural metal with fresh fluorescent red bands, and carried the ARDC fin flash. However, the flying display aircraft had the entire upper portion of the fuselage and the vertical tail painted in insignia red, with the fluorescent bands painted over it. The not normally-seen tow target storage box under the rear fuselage was clearly visible on this one.

North American T-28A, 52-1234. It was again a real surprise to see this older piston-engine trainer at Edwards, especially as it had not proved to be as effective a trainer as the USAF had planned. However, not publicly known at this time was that the USAF was beginning to develop the type as a counter-insurgency aircraft. This particular airframe may have been slated as part of that effort. What is certain, however, is that the aircraft is in the process of being repainted, with masking tape all over the fuselage markings.

McDonnell F4H-1F, 145310. This particular Phantom II was now in its third year at Edwards, and was being used for development of the various stores capable of being carried by the very large fighter. Shown are streamlined high-speed bombs, six being carried on a single outboard pylon, and three large unguided rocket carriers on the inboard pylon.

North American T-39A-1-NA, 59-2869. The second T-39A built, it was in extensive use at Edwards as a fast small transport. It is painted as previously, but there are now no visible unit insignia.

North American F-100D-1NA, 54-2127. The seventh F-100D built, shows the enlarged vertical tail, with a fairing for the fuel dump vent above the rudder. As the first of the new supersonic fighters, the F-100 ran into a lot of problems, and no less than 500 of them were lost to various causes between 1956 and 1970. As a result, by this time the USAF was planning to phase the F-100 out of service, but events in Vietnam caused the type to remain in service a lot longer than originally envisioned. This one is displaying the Edwards flight test center unit insignia.

North American F-100F-1-NA, 56-3726. This F-100F has been retained at Edwards since 1958 for testing, and eventual chase use. It shows the new anonymous style of markings.

(p. 130, Top Left p. 131) Northrop N-156F, 59-4987. The first F-5A "Freedom Fighter" was displayed carrying three French Nord AS 30 Maquette missiles under the wings and fuselage, together with Sidewinder missiles at the wing-tips. This was an unusual look at European missiles at Edwards, as normally only domestically produced weapons were tested there.

(See center) Northrop T-38A-NO, 59-1598. The T-38A trainer was now getting into large scale production, and here we see a stock version on display.

Northrop T-38A-NO, 59-1595. Another stock T-38A was used in the flying display, showing off its great maneuverability. Note that the service versions retained the overall white finish, with fluorescent bands. The white finish had originally been adopted as a method of finding fluid leaks, cracks, etc., quickly on the test aircraft, and proved so successful that it was adopted as standard.

1963

Nineteen sixty-three turned out to be an incredibly thin year at Edwards, with only two new aircraft on display, one being the Convair F-106B and the other being the huge commercial Aero Spacelines "Guppy" transport. At least the weather was fine! The aircraft that were displayed are listed below.

Manufacturer	Type	Popular Name	Aircraft Serial
North American	A3J-1	Vigilante	NASA WST 858
Boeing	NB-52NA-BO	Stratofortress	52-003
Boeing	NB-52NB-BO-10	Stratofortress	52-008
Martin	B-57E-MA	Canberra	55-4262
Martin	B-57E-MA	Canberra	55-4267
Convair	YB/RB-58A-CF	Hustler	55-0666
Convair	B-58A-CF	Hustler	61-2066
Douglas	B-66B-DL	Destroyer	53-0488
Lockheed	C-130A-LM	Hercules	53-3132
Lockheed	C-130A-LM	Hercules	53-3135
North American	F-100F-15-NA	Super Saber	56-3921
Convair	TF-102A-10-CO	Delta Dagger	54-1354
Lockheed	F-104A-10-LO	Starfighter	56-0750
Lockheed	F-104A-10-LO	Starfighter	56-0755
Republic	F-105D-31RE	Thunderchief	62-4309
Convair	F-106A-CO	Delta Dart	57-0238
Convair	F-106B-CO	Delta Dart	57-2519
McDonnell	F4H-1	Phantom II	145310
Vertol	HH-21B-VL	Workhorse	53-4361
North American	T-28A-NA	Trojan	49-1495
Lockheed	T-33A-1-LO	Shooting Star	52-9846
Lockheed	T-33A-1-LO	Shooting Star	53-5188
Northrop	T-38A-15-NO	Talon	59-1595
Northrop	T-38A-15-NO	Talon	59-1598
Lockheed	U-2D-LO		56-6701
Cessna	U-3A-CE	Blue Canoe	57-5847
Aero Spacelines	377PG	Guppy	N1024V

(Top, Right) McDonnell F4H-1F, 145310. The by now venerable F4H-1F test airframe was shown with Bullpup missiles under the inner pylons, 500 lb. bombs under the outboard pylons, and 750 lb. bombs under the fuselage pylon. All of this making one feel that the aircraft was still really an attack type (as originally intended), rather than the fighter designation it was being purchased under.

North American T-28A-NA, 49-1495. Beautifully finished, this T-28A is not the same one as that seen being repainted in the previous year. It is carrying lots of fluorescent red and shows the Systems Command marking on the lower fuselage side. Very unusual is the white finished upper portion of the fuselage, with the lower portion being finished in dark gray.

Boeing NB-52A-BO, 52-003. Now at Edwards for five years, the NB-52A took part in the flying display only, first accompanied by a Martin B-57, and a B-58, then on its own, prior to landing and taxiing past. No major changes were visible on its finish or markings.

(Top, Right) Boeing NB-52B-BO-10, 52-008. First time that the second test program B-52 had been displayed statically. Its markings differed slightly in application of the fluorescent orange aft fuselage bands, and in the shape of the fluorescent red nose areas. The X-15 was not displayed at all this year, either alone or with the B-52.

Lockheed T-33A-1-LO, 52-9846. Very nicely kept T-33A, with the new "O" for obsolete marked in front of the tail serial number. This was to be seen on many aircraft this year and later at Edwards; it denoted that the aircraft was ten years or more in age. Thus O-29846 meant that the aircraft was from a fiscal year 1952, rather than 1962, order. In later years, some aircraft survived so long in service that the "O" could even mean that the aircraft was twenty years or more old. Also displayed is the Systems Command badge.

(Top, Right) Douglas B-66B-DL, 53-0488. Again, we see one of the only 72 bomber versions of the B-66, but this was the first time that its sizable bomb-doors had been shown open. Missing from the rear of the fuselage is the original tail turret and its associated radome. It had probably been used for tests for some considerable time, as it has very faded fluorescent orange on the aft fuselage, together with the ARDC fin flash. However, the nose has been marked with the new Systems Command badge on very fresh fluorescent red paint.

Lockheed C-130A-LM, 53-3135. Apparently unchanged from its previous year's display appearance, this early C-130 was still being used for development work at Edwards.

Lockheed T-33A-5-LO, 53-5188. Unlike several of the other Fiscal 1953 aircraft present this day at Edwards, this T-33 in the flying display had the "O" added in front of its tail number.

Convair TF-102A-10-CO, 54-1354. Taking part in the flying display, the resident TF-102A showed a large new insignia on the vertical tail, together with the Systems Command badge on the left nose. Like all stock F-102s and F-106s, it is finished in gloss aircraft gray.

Convair YB/RB-58A-CF, 55-0666. Twenty of the original test batch of B-58As had been brought up to the current production status during 1961, but this was not one of them. In fact, it was still being used in various tests. It had flown at Mach 2 for some 32 minutes in 1958, used for tests at Holloman AFB, New Mexico, in 1959, and made the longest early program test flight of 11 hours, 15 minutes in 1962, and was now currently being used in subsonic low-level flights to ascertain the B-58's response in atmospheric turbulence. As such, it was completely devoid of any unit markings. The multiple small main landing gear wheels show up here: it was not uncommon for B-58s to blow several of the tires during takeoff (despite their 240 lb. pressure). The author saw this happen during a test at Edwards; it took over an hour to foam 10,000 ft of the runway. This was to prevent fire from developing from the metal wheels and other landing gear parts contacting the runway. The test pilot made a beautiful, smooth landing on the foam, no further damage being done to the aircraft (or runway). This particular aircraft eventually ended up being statically displayed at Chanute AFB, Illinois, with the bogus serial number 61-2059.

Lockheed C-130A-LM, 53-3132. Only difference in the flying display C-130A was removal of the buzz-number from the fuselage.

Martin B-57E-MA, 55-4262. Again no differences visible in the flying display B-57E; it is still carrying the large raked aft antenna on the top of the rear fuselage, purpose unknown. The static display aircraft (55-4267) appeared to be identical.

(See also Top p. 139) Lockheed F-104A-10-LO, 56-0750, 56-0755. Two of the development batch seen previously at Edwards, these F-104As showed no major changes. The flying display aircraft did not have its wings painted white, unlike its static display companion.

North American F-100F-15-NA, 56-3921. From the first batch of F-100Fs built, this aircraft taxiing past showed not a single distinguishing mark or badge, so it is hard to say what its purpose was, other than giving its pilot some flying time!

Convair F-106B-CO, 57-2519. First time seen at Edwards was the two-seat version of the F-106, the F-106B. Like the F-106A in the flying display, it carried the Systems Command badge on the nose, and also had the entire rudder painted in red. This was the thirteenth F-106B of the 63 built. This version retained the full operational capability of the single-seat aircraft, and they served with every squadron equipped with the F-106A.

Cessna U-3A-CE, 57-5847. Now becoming a fixture at these open houses, this U-3A retained a modified version of the original blue and white color scheme. The right door of the right engine cowling has been left open to let us see that it actually had piston engines in this day and age of jets!

(Right, Bottom) Lockheed U-2A-LO, 56-6701. This aircraft looked just like it had during the two previous years, but contrary to most of the other aircraft at Edwards, it had some additional fluorescent patches added to the vertical and the wing tip skids.

Convair F-106A-CO, 57-0238. The F-106A also appeared to be just another standard airframe, but at least it did have the Systems Command badge on the nose.

Convair B-58A-CF, 61-2066. A standard production version of the B-58A took part in the fly-past and the taxi-by. It was another totally anonymous aircraft, being the fifteenth from last of the production run. After being used for test work by Convair, it went to the 43rd Bomb Wing at Carswell AFB.

Northrop T-38A-15-NO, 59-1598. The static display T-38A showed a change to the color scheme used on the earlier versions. Fluorescent red was used on the nose, and now on the vertical tail, rather than the rear fuselage. Four T-38As took part in the fly-past: two had the original paint scheme (somewhat faded), and the other pair had the newer paint scheme. Incidentally, the new scheme did show up better in the fly-past.

Republic F-105D-31RE, 62-4309. From the last batch of F-105Ds built, this aircraft showed the new aluminized overall finish being applied to all F-105s in service. It was primarily applied to help prevent corrosion, which was becoming a problem with F-105s in Europe and the Pacific. Once again, a total lack of unit insignia.

North American A3J-1, NASA WST 858. This A3J-1 is BuNo. 147858, but exactly why NASA was involved with it is a mystery. The inscription WST 858 on the nose may stand for Weapon System Test, as the Navy had a great deal of trouble in trying to get the unique rear-ejecting bomb-bay to work. In any event, it was rendered unnecessary when the nuclear attack role was removed from the Navy's missions.

Aero Spacelines Guppy, N1024V. The only new type displayed at Edwards in 1963 was the astounding Aero Spacelines Guppy. Greatly modified from a Boeing Stratocruiser, the Guppy had a huge, enlarged upper fuselage, intended to carry outsize cargo of great bulk, but light weight. Designed by Jack Conroy at Santa Barbara, the Guppy caused intense interest, plus lots of disbelief, wherever it was seen. It has been related that when the Guppy was sent later to St. Louis to pick up the prototype McDonnell Mercury space capsule, it was parked on the ramp outside Mr. McDonnell's office. It is alleged that when he saw it, that he refused to allow the capsule to be loaded onto such an unsafe vehicle! In fact, Northrop was given a crash program to verify the flying capabilities of the Guppy, supposedly as a result of Mr. Mac's views. Nevertheless, the Guppy went into widespread usage, and spawned off a number of later versions. This eventually culminated in the Airbus re-design of an A300 into a similar vehicle, named the "Beluga."

Aero Spacelines

Spacelines

Aero
LARGEST AIRPLANE
IN THE WORLD

1964

In contrast to the previous year, there were more aircraft displayed at Edwards in 1964, including some refreshing new types. These included the Douglas A-4E, Douglas C-118A, Grumman OV-1B, Lockheed P-3A, McDonnell F-4B and F-4C, North American YAT-28E, Northrop F-5A and F-5B, and the Ryan XV-5A. The new DoD aircraft designation system adopted in 1963 began to show up on the aircraft. Also apparent was the almost complete disappearance of fluorescent paint on the aircraft, its use now being confined to types previously displayed or based at Edwards. The Ryan XV-5A was yet another step toward a usable VTOL/STOL aircraft, but its unusual use of fans in the wings to generate vertical lift did not result in an operational type. The aircraft that were displayed this year are listed below.

Manufacturer	Type	Popular Name	Aircraft Serial
Douglas	A-4E	Skyhawk	149657 from VX-5
Boeing	NB-52A-BO	Stratofortress	52-003
Martin	B-57E-MA	Canberra	55-4258
Martin	NB-57E-MA	Canberra	55-4267
Convair	YB/RB-58A-CF	Hustler	55-0666
Convair	B-58A-CF	Hustler	61-2066
Douglas	B-66B-DL	Destroyer	53-0488
Douglas	C-118A-DO	Liftmaster	51-3826
Lockheed	JC-130A-LM	Hercules	54-1624
Lockheed	C-130A-LM	Hercules	54-1626
McDonnell	F-4B	Phantom II	151438 from VF-96
McDonnell	F-4C-15-MC	Phantom II	63-7410
McDonnell	F-4C-17-MC	Phantom II	63-744
Northrop	F-5A-15-NO	Freedom Fighter	63-8370
Northrop	NF-5A-15-NO	Freedom Fighter	63-8369
Northrop	F-5B-5-NO	Freedom Fighter	63-8438
McDonnell	JF-101A-1-MC	Voodoo	53-2422
Convair	TF-102A-5-CO	Delta Dagger	54-1354
Lockheed	F-104A-10-LO	Starfighter	56-0754
Lockheed	F-104A-10-LO	Starfighter	56-0755
Lockheed	NF-104A-10-LO	Starfighter	56-0756
Lockheed	F-104D-5-LO	Starfighter	57-1316
Convair	F-106A-CO	Delta Dart	57-238
Convair	F-106B-CO	Delta Dart	57-2519
Vertol	JHH-21B-VL	Workhorse	53-4361
Lockheed	P-3A	Orion	150529
North American	P-51D	Mustang	N2251D
North American	YAT-28E	Trojan	0-13786
Lockheed	T-33A-1-LO	Shooting Star	53-4999
Lockheed	T-33A-5-LO	Shooting Star	53-5188
Cessna	T-37A-CE	Tweet	56-3492
Northrop	T-38A-15-NO	Talon	59-1597
Northrop	T-38A-15-NO	Talon	59-1600
Northrop	T-38A-15-NO	Talon	59-1601
Northrop	T-38A-50-NO	Talon	63-8135
North American	T-39A-1-NA	Saberliner	59-2869
Lockheed	U-2A-LO		56-6722
Lockheed	U-2D-LO		56-6721
Cessna	U-3A-CE	Blue Canoe	57-5849
Grumman	OV-1B-GR	Mohawk	62-5873
Ryan	XV-5A		62-4506

(See also Top, Center p. 148) North American YAT-28E, 0-13786. It was strange to see a new aircraft variant for the first time with the "O" prefix in front of the serial number. However, this very interesting new aircraft, the YAT-28E, was a conversion from a production T-28A, fitted with a Lycoming T55 turboprop in place of the big Wright R-1300 piston engine. The aircraft had been heavily modified for ground attack use: the modifications included the addition of two 0.5 in guns, one under each wing, and three stores pylons outboard of the guns, under each wing. The vertical tail had its height increased, while the engine had a four-bladed propeller. The color scheme was also very unusual, a gloss dark green on upper surfaces, and gloss white lower surfaces. Much attention was being paid at this time to finding cheaper aircraft to carry out the so-called "counter insurgency" operations, needed all over the world, as part of the on-going strife between the Western nations and the Soviet Union. The YAT-28E testing resulted in a fatal crash, and it was not selected for production (however, some of its systems found their way into even-cheaper, simpler, conversions of some 393 T-28As into T-28Ds and AT-28Ds, for the same role).

Vertol JHH-21B-VL, 53-4361. The long-time standby rescue aircraft at Edwards remained the ubiquitous H-21B, looking just like it had for several years.

(Top, Right) Douglas A-4E, 149657. VX-5. Following the major aircraft re-numbering changes ordered by Defense Secretary McNamara in 1962, the Douglas A4D had now become the A-4, and the "E" version was originally the A4D-5. Main change from previous versions was the fitting of the later Pratt & Whitney J52 engine, in place of the by then out of production Wright J65. The increased thrust, together with decreased weight and fuel consumption, allowed the addition of two extra stores stations to the existing three under the wings and fuselage. Improved bombing and navigation systems also improved the capability of this new Skyhawk version. No less than 500 were eventually delivered. This one is seen in the standard Navy gray and white scheme, and carries the light green band of VX-5, the test unit at nearby NAS China Lake.

Lockheed T-33A-5-LO, 53-5188. Same as before, this essential training aircraft also looked exactly as in previous displays.

McDonnell F-4B, 151438, VF-96. For the first time, we saw production versions of the mighty F4H-1, now re-designated F-4 (the earlier F4H-1Fs had been re-designated as F-4As, under the new system). Not only that, this was an operational F-4B, from VF-96, at that time based on the *USS Ranger* carrier. Very obvious changes included the new, larger radome, with an infrared sensor below, a larger cockpit, giving more room for the rear seat, and much larger intakes for the more powerful J79-GE-8 engines. Two of the F-4Bs were delivered to the USAF in 1962 for evaluation, designated by them as F-110s. These tests, together with considerable pressure from McNamara, caused USAF to designate the F-110 as the standard fighter and tactical reconnaissance aircraft for TAC, USAFE, and PACAF. In fact, it was the two designations (F4H-1 and F-110A) for this aircraft that caused McNamara to lose his temper in a heated Congressional inquiry. Not being able to remember both designations at the time, in a fit of pique he ordered that all aircraft of the same type in any of the three services should have the same designation. Regrettably, this wrecked the very logical designation systems then in use by the USAF and Navy, and replaced them with a single system of no logical form whatsover. No-one has ever been able to calculate the cost of this sweeping decision, but it probably ran into millions of dollars over the next several years, incurred by having to repaint aircraft designations en masse, change all manuals relating to each type, etc.

Lockheed JC-130A-LM, 54-1624. Continuing its test work at Edwards, this JC-130A appears to be unchanged from its previous displays.

Douglas C-118A-DO, 51-3826. Carrying a very small serial number on the vertical tail, this C-118 presumably replaced the earlier, un-pressurized, C-54A transport at Edwards. There are traces of a previous marking band across the tail, possibly from MATS. As seen, it was in natural metal finish all over, without even a white top to reduce the cabin heat. Most likely, it just arrived at Edwards, and had not yet been repainted (checking on the C-54/C-118s displayed later at Edwards finds no trace of this particular one, so its use remains a mystery).

Boeing NB-52A-BO, 52-003. Old faithful was only seen in the flying display this year, and the other NB-52 was not visible. The X-15 was also conspicuous by its absence, although both the first and third aircraft were making regular flights. Probable reason was that NASA was only a few weeks away from the first test flight of the second X-15. This had been rebuilt as the X-15A-2, after its earlier ground mishap.

Douglas B-66B-DL, 53-0488. This year the B-66B had no extra paint on it (except for the white top behind the cockpit canopy), and no visible unit markings. Otherwise, it appeared to be similar to its previous state.

ConvairYB/RB-58A-CF, 55-0666. Only unit marking visible on the B-66 was the ARDC badge aft of the radome. Despite the "triple 6" number connotation in some areas, this aircraft survived all of its testing for various systems on the type, and finally retired to Chanute AFB, Illinois, but carrying the spurious serial number 61-2059.

Martin NB-57E-MA, 55-4267. This year the displayed B-57E was in extremely plain natural metal finish, showing only its "buzz-number."

Lockheed F-104A-10-LO, 56-0755. New paint had finally been added to this long-time resident at Edwards, in the form of light gray on the vertical tail, and on the bottom half of the fuselage forward of the aft fuselage. The wings were also painted in the now standard white finish.

Lockheed NF-104A 10-LO, 56-0756. We had seen this aircraft before, but this year it had been modified to add a 6,000 lb. thrust rocket motor above the jetpipe. Also fitted were the enlarged vertical tail surfaces of the F-104G version, the wing tips were extended 2 ft each, and hydrogen-peroxide control thrusters were added at the nose, wingtips, and tail. All of these modifications were made to three F-104As, this one, plus 56-760 and 56-762. They had been converted for use in the new USAF astronaut training program conducted at the Aerospace Research Pilot School at Edwards AFB. The first one was demonstrated in the flying display this year, and showed an astonishing rate of climb on its combined power (not to mention the incredible noise!). The rocket motor showed the usual highly visible exhaust when lit just after lift-off.

Lockheed U-2A-LO, 56-6722. A new U-2 was shown in the flying display this year, this time in a chiefly black (with large white areas under the wings, horizontal tail, and fuselage belly) paint scheme, without any unit markings. It was one of the earlier Edwards AFB based U-2Ds rebuilt as a U-2A for the HICAT (high-altitude clear air turbulence) program, and was equipped with a red and white barber-pole striped probe under the nose. The HICAT program lasted for several years, taking the aircraft to Fiji, New Zealand, Australia, and Britain. This aircraft eventually went to the USAF Museum, Dayton, Ohio.

(Top, Right) Lockheed U-2D-LO, 56-6721. "Smokey Joe" was back on the vertical tail of the U-2D this year (however, it was now painted below the serial number). A new research device was apparent in the form of a large diameter white ball mounted on top of the fuselage, between the two cockpits. The ball had a large flat optical panel on its forward face, as shown (it was probably rotatable).

Convair F-106A-CO, 57-238. Another stock F-106A assigned to the Systems Command, no apparent changes visible.

Cessna U-3A-CE, 57-5849. This year, the "Blue Canoe" was seen with a new white top to the fuselage, white trim on the vertical, and the fluorescent tip tank colors also changed to white.

Northrop T-38A-15-NO, 59-1601. No changes were visible on this stock T-38A assigned to Systems Command, but the paint scheme had now changed to be just an overall white, without any red paint or fluorescent areas.

Grumman OV-1B-GR, 62-5873. This was a new type not previously seen at Edwards, and it was only displayed in the flying display. It originated as long ago as 1956, from a joint Army-Marine Corps requirement. The Marines dropped out before completion of the first prototype, designated YAO-1. This made its first flight in April 1959, powered by two Lycoming T53 turboprops. Designed as a battlefield surveillance aircraft, some 287 of the production OV-1A, B, and C versions served in Vietnam. An improved version, the OV-1D, incorporated all of the various capabilities of the three earlier versions, from Vietnam experience. Grumman was not normally associated with land-based aircraft, but the Mohawks gave good service with the Army. Its late appearance at Edwards, finished in the usual Army olive drab, is a remainder that the Army conducted most of its flight testing elsewhere in the USA.

Ryan XV-5A, 62-4506. This was a surprising exhibit, as interest in research vertical takeoff aircraft had diminished since all of the earlier types displayed in 1959. However, the new Ryan XV-5A used a different technique, that of horizontal fans embedded in the wing surfaces. It had originally been designated the VZ-11, but this had changed in the 1962 tri-service changes discussed earlier. It had not yet made its first flight, that taking place on May 25, 1964, nine days after its public showing here. Powered by two GE J85 jets, two were built, this being the second one. It is finished in the usual Army gloss olive drab finish.

624

U.S.ARMY

XV-5A

McDonnell F-4C-15-MC, 63-7410. The fourth production F-4C, of some 583 built, is seen in the flyby and taxiing past. It still carries the infrared housing below the nose radome, though it remained empty.

McDonnell F-4C-17-MC, 63-7445. This is the 39th F-4C for the USAF, and shows off some of its huge stores carrying capability. Note that these first production versions of the F-4C retained the complete Navy paint scheme: in fact, they were often referred to as "bastardized" Navy aircraft by USAF personnel. USAF was not at this time too happy with McNamara having crammed the F-4 down their throat, in preference to their own F-105 and F-106 fighters. Appreciation of its capabilities did not really come until it had been committed to combat in Vietnam, and the USAF did not get a version fully meeting their needs until the F-4D was put into production.

Northrop NF-5A-NO-15, 63-8369. This is the third production F-5A, parked with a very large display of all of the sundry stores and nasty weapons it could carry. Flight testing had proceeded well at Edwards, but the type was still regarded by the fighter pilots as more of a toy than a really serious fighter. In the initial round of selling in Europe, it had been severely beaten by the upgraded Lockheed F-104G. However, after a few years of operating F-104Gs, those air forces were finding out its severe operating costs, both in money and in lost pilots. As a result, many of them ultimately ended up adding F-5s to their inventory. Thus, more than 1,100 F-5As and F-5Bs were built by 1972.

1965

With the war in Vietnam growing into a full-scale conflict, there were now more aircraft being tested at Edwards, and this also resulted in more new aircraft being displayed. This year's crop was much larger than we had seen for many years and included the following; Beech U-8, Boeing KC-97H, Lockheed YF-12, Lockheed C-141, Lockheed HC-130H, McDonnell RF-4C, North American XB-70, Northrop X-21, and the Republic F-105F. Although not all of these were new aircraft, they were types that we had not seen at the previous displays. All of the aircraft displayed are listed below.

Manufacturer	Type	Popular Name	Aircraft Serial
Douglas	A-3D-2	Skywarrior	142241 VAH-4
Boeing	NB-52A-BO	Stratofortress	52-0003
Martin	NB-57A	Canberra	52-1550
Douglas	B-66B-DL	Destroyer	54-0481
North American	XB-70A-NA	Valkyrie	62-0001
Douglas	C-54M	Skymaster	44-9093
Boeing	KC-97H-BO	Stratofreighter	53-0315
Lockheed	JC-130A-LO	Hercules	53-3132
Lockheed	C-130E-LM	Hercules	64-0560
Lockheed	HC-130H-LM	Hercules	64-14853 MATS
Lockheed	C-141A-LM	Starlifter	61-2779
Lockheed	C-141A-LM	Starlifter	63-8075
McDonnell	F-4C-15-MC	Phantom II	63-7407
McDonnell	RF-4C-18-MC	Phantom II	63-7743
Northrop	F-5A-15-NO	Freedom Fighter	63-8371
Chance Vought	F-8C	CrusaderII	146995 VU-7
Lockheed	YF-12A-LO	Blackbird	60-6935
Convair	TF-102A-10-CO	Delta Dagger	54-1354
Lockheed	F-104A-10-LO	Starfighter	56-0748
Republic	F-105F-1-RE	Thunderchief	63-8263
Convair	F-106A-CO	Delta Dart	57-0238
General Dynamics	F-111A-CF	Ardvark	63-9767
Bell	UH-1B-BF	Iroquois	63-8684
Vertol	JCH-21B-VL	Workhorse	51-15854
Lockheed	T-33A-15-LO	Shooting Star	53-5101
Northrop	T-38A-15-NO	Talon	59-1600
Lockheed	U-2D-LO		56-6721
Lockheed	U-2A-LO		56-6722
Cessna	U-3A-CE	Blue Canoe	57-5849
De Havilland	U-6A-DH	Beaver	58-2039
Beech	U-8D-BH	Seminole	58-3090
Ryan	XV-5A		62-4505
North American	X-15		56-6670
Northrop	X-21A		55-0408

Douglas A3D-2, 142241, VAH-4. One of the first production A3D-2s, this had a beefed up airframe, stressed for a load factor of 3.4g versus the earlier 2.67g, and more powerful J57 engines. The first 123 of this version did not have the refueling probe fitted to later A3Ds (but most of them had it fitted later). The type had been re-designated as the A-3B by the new DoD numbering system, but this VAH-4 one had not yet had its fuselage designation changed.

Vought F-8C, 146995, VU-7. Third service version of the Crusader, the F-8C was produced for the air superiority role, with increased thrust from the Pratt & Whitney J57 engine. Also carrying the small nose-mounted radar for improved all-weather capabilities, introduced on the F-8B version, the F-8C could be distinguished by the two ventral fins added to improve directional stability. However, the type was restricted to Mach 1.7 by other stability problems. The F-8 gained a reputation of being a real dogfighter in Vietnam, sometimes attributed to its internal gun armament. However, the greater number of its kills were made with its Sidewinder missiles. Due to a general lack of newer aircraft being tested at Edwards AFB, the USAF test center was now inviting service units to the open house, hence the operational Navy and USAF types on display.

Douglas C-54M, 44-9093. In contrast to the previous year, the perennial C-54 was present once again, this year's version carrying a very neat white top, and trim aft of the cockpit. It was really surprising to see the old Skymaster still in service twenty years after the end of WWII. This was one of thirty converted for carrying 30 stretchers, plus medical attendants, during the Korean War. Note that the cabin windows have been converted from the normal round C-54 type to the rectangular ones used on the C-118s.

Vertol JCH-21B-VL, 51-15854. Now carrying its new DoD designation, the Workhorse looked the same, still carrying out its essential search and rescue duties for the huge base.

Boeing NB-52A-BO, 52-0003. Now in its seventh year at Edwards, the NB-52A was displayed with the first X-15, 56-6670, mounted under its right wing. The large number of flight symbols painted on the B-52's forward right fuselage was testimony to the tremendous amount of research flights being made on the X-15 program. By this time, no fewer than 129 flights had been made, and all three X-15s had been fitted with the large XLR99 rocket motor, giving over 57,000 lb. of thrust at sea level.

Martin NB-57B-MA, 52-1550. Just like all of the other B-57s seen at Edwards over the years, there was no indication of any particular test or role to be seen. However, the fluorescent paint days were definitely over, and this early B-57B is in very clean natural metal finish all over. Note how the now removed "Buzz Number" has etched itself into the aft fuselage skin.

Boeing KC-97L-BO, 53-0315. It was a surprise to see the old Stratofreighter tanker again, but this was a new 1964 version. Eighty-two of them were fitted with J47 jet pods removed from surplus KB-50s, and assigned to ANG units. This one was from the Tennessee ANG, in a shiny natural metal finish.

Lockheed JC-130A-LO, 53-3132. Another old-timer at Edwards was the C-130A, now in its ninth year at the open house. It had also lost all of its various red and fluorescent paint finishes, along with the "buzz numbers." Lacking the later nose radome, it showed the ARDC badge up front. A change in policy now allowed the public access to the interior of the fuselage, via the open rear ramp.

Lockheed T-33A-15-LO, 53-5101. This year's T-33 from the AFSC had also lost all of its paint, even the tip-tanks were now polished natural metal.

Douglas B-66B-DL, 54-0481. Taking part only in the flying display, the B-66 was also in natural metal overall (like it was in 1957), but this one carried the cryptic letters "ZD" on the aft fuselage speed brakes, together with a unit citation award on the vertical tail. It was not known at the time, but the type was not to be seen again at Edwards for another eight or nine years.

Convair TF-102A-10-CO, 54-1354. This long time resident at the Edwards Test Pilot School showed no visible changes from previous displays.

Northrop X-21A, 55-0408. A brand new "X" aircraft was now a rare event, but the X-21A was also very different from most of the earlier X-types. A very heavily modified B-66 airframe, it had been fitted with a totally new wing, and had new engines, moved from the wing position to the aft fuselage. Two of the X-21s were built, this being the first one. The wing had been built to extraordinarily tight tolerances, and had its surface cut by extremely thin slots. All of this was to try and obtain laminar flow over the entire wing; in theory this would decrease drag so that the aircraft's range would be increased by about 50%. The first X-21 had made its first flight from Hawthorne to Edwards in April 1963, but it had not been seen outside on static display, having been kept in the large hangar during the 1964 open house. In fact, the program had been terminated in that year, because it had been found that maintenance of such wings would have been impractical in normal use. There were two chief problems, (1) dust, dirt, and insects impinging on the surface caused lack of laminar flow, and would have taken tremendous efforts to keep the surfaces clean, and (2) moisture sucked into the wing's internal structure caused internal corrosion, in areas that could not be inspected or corrected for structural degeneration. The X-21's natural metal wings contrasted strongly with the white fuselage, vertical tail, and engine pods, trimmed with fluorescent orange, and touched up with fluorescent red.

Lockheed F-104A-10-LO, 56-0748. This early F-104A now no longer carried any of the Edwards unit badges, and freshly painted with aircraft gray, may have been prepared for use elsewhere.

Lockheed U-2D-LO, 56-6721. There were no visible differences in the U-2D this year at Edwards, but at least, our supposition, that the white ball above the cockpits could rotate, was proved correct! (note its position, as compared to the previous year).

Lockheed U-2A-LO, 56-6722. Also, there were no visible changes to the other resident Edwards U-2A, which again only took part in the flying display.

Convair F-106A-CO, 57-0238. Some of the later Century series fighters spent a lot longer at Edwards than originally planned, and this F-106A was one of them. This year, the Air Force funded several modernization programs to the F-106A, by then regarded as its best interceptor. These included a new TACAN system, and the fitting of new supersonic under-wing fuel tanks. The changes were made as a stop-gap measure, while the USAF was studying new advanced manned interceptors, such as operational versions of the YF-12A and F-111.

Cessna U-3A-CE, 57-5849. This U-3A was by now a perennial at the open house, but this year its wing tip tanks had been painted black, with the old AFFTC white lightning bolt along them (minor protest against the by now universal USAF anonymity?).

De Havilland U-6A-DH, 58-2039. The Beaver was back in a very glossy olive drab Army finish, with fluorescent red panels on the nose, aft fuselage, wings, and tail.

Northrop T-38A-15-NO, 59-1600. Still carrying "buzz numbers," the resident T-38A also shows the lack of any color other than its base white finish.

(Top, Right) Beech U-8D-BH, 58-3090. Formerly the L-23, the U-8D fleet consisted of 85 new aircraft delivered in 1956-1958, and a further 93 L-23As and L-23Bs rebuilt as L-23Ds, these having new serial numbers. The very clean U-8D shown at Edwards was one of the last of the rebuilt aircraft (there were only two more after this one), and looks as good as new. They were used as six-seat Army staff transports. It is also in the same olive drab finish as the Beaver, but does not have any fluorescent trim. However, it does have a very neatly trimmed white top to the fuselage.

Lockheed YF-12A-LO, 60-6935. At last, we were to see the majestic YF-12A triple-sonic twin-jet interceptor, although only in the flying display. It looked very sinister in its all black finish, with just a few white markings. We were going to have to wait another year before seeing the type in the static display. It had made its first flight in July 1962, but it was not heard of publicly until President Johnson announced its existence in February 1964. The first display to the press was at Edwards in September 1964, but its appearance at this year's open house was the first time the general public had been able to see the YF-12A. The long slender fuselage had chines similar to the X-15, but the delta wing with two large outboard jets was a new feature for a USAF type.

Lockheed C-141A-LM, 61-2779. As the first major four-jet transport for the Military Air Transport System, the C-141A aroused a great deal of interest. Winner of a major contest between Boeing, Convair, Douglas, and Lockheed, the type had originated as a rather straight-forward jet version of the C-130, and used the same fuselage cross-section. Demand for a higher performance led to the wings being swept moderately, and a change to four Pratt & Whitney TF33 turbofans conferred good performance with a reasonable range. The first one had flown in December 1963, with deliveries to MATS beginning in October 1964. Operational use was begun in April 1965, just before this display. Later this year, the C-141A began regular flights to Vietnam. Some fourteen squadrons were finally equipped with the C-141A, and it became the mainstay of MATS long range operations worldwide. Very noticeable as it taxied past were the two large rear clam-shell doors. The C-141A suffered from two main problems in service; (1) the narrow width of the fuselage made it difficult to drive most Army vehicles into it backwards, thus taking a considerable amount of time to fully load the aircraft (this was done to allow the vehicles to drive out in a forward direction at the destination, speeding up the unloading), (2) structural fatigue caused considerable rework, and reduction in availability during the 1980-1990s time frame. This aircraft was the fifth one built, resplendent in its natural metal finish.

North American XB-70A-NA, 62-0001. Again, this was the first time a major new aircraft had been seen at Edwards, the huge triple-sonic Valkyrie bomber, which had made its first flight in September 1964, from Palmdale to Edwards. Unfortunately, demonstrated Soviet SAM (surface-to-air missile) capability had rendered its mission untenable, so only the two aircraft were built. After some minor early problems, the first one provided a great deal of high speed research, and is at time of writing the only large aircraft to have flown consistently at Mach 3. It was an awesome sight, with its long slender nose in front of the huge delta wing. Under the wing was a large tapered box containing no less than six full-time afterburning J93 engines. This first prototype was eventually flown to the USAF museum at Dayton, Ohio, and can be seen there to this day. The B-70 and the YF-12/SR-71 aircraft have more sustained flight time at Mach 3 plus than all other types world-wide.

(Left, Bottom) Ryan XV-5A, 62-4505. This is the first of the two XV-5As, and showed no obvious differences from the second one shown in the previous year. It was noticeable that none of these experimental VTOL aircraft were ever flown publicly.

(Right, Center) McDonnell F-4C-15-MC, 63-7407. By now production of the F-4 for the USAF was under full swing, and production model F-4Cs were showing up at Edwards for a full range of testing. Some 583 were built for TAC units, and this is the very first production F-4C for them. Most of the changes made for USAF use were internal, such as dual controls for the crew, an inertial-navigation system, flying-boom refueling (replacing the Navy probe and drogue system), and cartridge-start for the J79 engines. Externally, the paint scheme was the same as the Navy version, with USAF markings. However, a much improved version, the F-4D, was already in production in 1965 for USAF units. The F-4C had already gone to war in Vietnam at the start of this year, but no one realized that this was to be the longest war the USA had been involved in to date. One result was a huge increase in testing activity at Edwards AFB, which became very apparent in 1966 (see next Chapter).

McDonnell RF-4C-18-MC, 63-7743. A badly needed replacement for the RF-101s became available in 1963, in the initial form of two YRF-4Cs (modified Navy F-4Bs). These had a 33 inch longer nose, filled with cameras; side-looking radar and an infra-red scanner were fitted in the fuselage. The four fuselage-mounted Sparrow missiles and their associated recesses were deleted. First production version flew in May 1964, and this is the fourth aircraft off the line. The type had a long service life, continuing in squadron service into the late 1980s.

(Left, Center) Lockheed C-141A-LM 63-8075. This is the first of the 1963 contract C-141s, and has marked on both sides of its nose "2500 hours in 11 months," together with other cartoons and markings, indicating the extent of the flight testing prior to MATS service. It carries the MATS band across the vertical tail. Unusually, we see it taxiing both ways on the ramp (normally, aircraft taxied back on the ramp in an easterly direction, so we usually saw only their left sides).

Republic F-105F-1-RE, 63-8263. The two-seat version of the Thunderchief seen here displayed the new TAC badge on its taller vertical tail, together with the 31-inch longer fuselage to accommodate the second cockpit. The F-105Fs were built by converting the last 143 F-105Ds in production. Losses of the F-105D single-seat version had begun to mount this year over North Vietnam, some 350 being lost between January 1965 and November 1968. The F-105Fs began to be modified later this year, into the "Wild Weasel" version, for hunting down and destroying radar and SAM sites, with the first 13 going to Vietnam in mid-1966. A few F-105Fs, dubbed Combat Martins, were modified with jamming equipment to block communications between the Vietnamese MiGs and their ground controllers. Others carried out extremely hazardous night, all-weather, low-level radar bombing missions, under the title of Commando Nail.

(Top, Right) Northrop F-5A-15-NO, 63-8371. Weapon testing of the F-5A continued at Edwards, and here we see a Bullpup missile, together with low-drag bombs mounted under the wing and fuselage.

Lockheed C-130E-LM, 64-0560. This year's C-130 was a late production version of the Hercules, which had been ordered in 1961 as an expedited effort to modernize MATS (now MAC) airlift capability. Range had been increased by the addition of large external tanks mounted between the engines on both wings, and the aircraft's permissible maximum weight had been increased. Almost 500 of this version were built by Lockheed, 377 of these going to MATS and TAC. The remainder went to foreign air forces, the Navy, Coast Guard, and the FAA. This example is a squadron service one, displaying a unit citation badge across the tail.

(Right, Center) Bell UH-1B-BF, 63-8684. The "Iroquois" appeared in a plain overall olive drab finish, but it was still a gloss one. Large numbers of the UH-1B went to Vietnam in the utility role, this version having nine seats in the cabin.

Lockheed HC-130H-LM, 64-14853, MATS. Also displayed for the first time, was the second HC-130H, a specialized rescue version of the C-130H. The C-130H was a new version, appearing in March 1965, and was an improved "E," with upgraded engines and sundry system improvements; it became the standard version of the Hercules, replacing the "E." The HC-130H was a special version developed for the Aerospace Rescue and Recovery Service, the first one flying in November 1964. It was fitted with the nose-mounted Fulton yoke equipment, 43 being built. Not visible in this view is the re-entry tracking radome, on the top of the forward fuselage.

General Dynamics F-111A-CF, 63-9767. This was the first time the very controversial F-111 had been displayed. It resulted from McNamara's desire to use one type of universal aircraft for both USAF and Navy combat roles. However, the weight of the Navy version finally grew to the point at which it would only be able to operate from a handful of the largest carriers. Also, the selection by McNamara of General Dynamics as the contractor was also controversial, because it over-ruled the Air Force selection of Boeing. The F-111A's first flight had been made on December 21, 1964, while the second of two prototypes made its first flight in February 1965, this being the aircraft displayed both statically and in the flying display at this year's open house. Engine problems and weight increases soon became major problems plaguing the program. Eventually the F-111A's final takeoff weight at 92,000 lb. exceeded the OSD specification by no less than 30,000 lb. (though only 10,000 lb. more than the Air Force had expected; this shows the major difference between politically motivated design versus mature service design requirements). The F-111A seen here was in natural metal finish, showing only the service name, national insignia, and serial number. Note the unusual full-span flaps, with the red-painted danger-warning finish in front of them. Much, much more was to be heard of this program for many years, as we shall see.

U.S. AIR FORCE

U.S. AIR FORCE

Douglas C-133A, 57-1613, MATS. Taxiing in is the 33rd C-133A, seen in standard MATS finish and markings, as opposed to the test versions we had become accustomed to seeing at Edwards. It is assigned to the 84th Air Transport Squadron, but apparently has been sent to Systems Command temporarily for some extra testing. Fifteen C-133Bs, with increased weight and power, followed the C-133As, the first being delivered to Travis AFB, California, in March 1960. Despite its service handling problems, the C-133 was the only transport capable of carrying all of the current service IRBMs and ICBMs, and as a result it remained in service until the first C-5s were available in the early 1970s.

1966

This is the last year to be covered in this volume, and it turned out to be the biggest display of them all, with no less than seven new types to be seen. However, it should be noted that two of these types were U.S. Navy aircraft; the new aircraft were the Grumman A-6A, Grumman E-2A, Douglas TA-4F, Northrop HL-10, Boeing-Vertol JCH-47A, Sikorsky JCH-3C, and the remarkable LTV/Hiller/Ryan XC-142A. Also shown only in the fly-pasts was the second North American XB-70A.

All of this activity was a sign of the full-scale war now raging in Vietnam, and one could not help noticing the growing number of helicopters entering service, and the marked decrease in new bomber aircraft. This was a result of the

Manufacturer	Type	Popular Name	Aircraft Serial
Douglas	A-3B	Skywarrior	147650 VAH-2
Douglas	A-4C	Skyhawk	145111 VA-194
Douglas	A-4C	Skyhawk	149626 VA-195
Douglas	TA-4F	Skyhawk	152848
Grumman	A-6A	Intruder	149480 NATC
Boeing	NB-52B-BO-10	Stratofortress	52-008
Martin	NB-57E-MA	Canberra	55-4262
North American	XB-70A-NA	Valkyrie	62-001
North American	XB-70A-NA	Valkyrie	62-207
Douglas	C-54M	Skymaster	44-9093
Douglas	C-118A-DO	Liftmaster	53-3265
Lockheed	C-130B-05-LM	Hercules	57-525
Lockheed	HC-130H-105-LM	Hercules	64-14825
LTV/Hiller/Ryan	XC-142A		62-5921
Grumman	E-2A	Hawkeye	151707 VAW-11
McDonnell	F-4B	Phantom II	152992 VF-121
McDonnell	F-4C-15-MC	Phantom II	62-12199
Northrop	F-5A-15-NO	Freedom Fighter	63-8421
Chance Vought	F-8E	Crusader	150672
Lockheed	YF-12A-LO	Blackbird	60-6935
Lockheed	YF-12A-LO	Blackbird	60-6936
Lockheed	F-104A-10-LO	Starfighter	56-755
Republic	F-105F-1-RE	Thunderchief	62-4445
General Dynamics	F-111A-CF	Ardvark	63-9773
General Dynamics	F-111A-CF	Ardvark	63-9774
Bell	UH-1D-BF	Iroquois	60-6069
Sikorsky	JCH-3C-SI	Jolly Green Giant	62-12577
Bell	OH-13E-BF	Sioux	51-13793
Vertol	JCH-21B-VL	Workhorse	53-4389
Hiller	OH-23D-UH	Raven	55-3062
Boeing-Vertol	JCH-47A-BV	Chinook	60-3449
Northrop	HL-10		NASA 804
Lockheed	P-3A	Orion	152617
Lockheed	T-33A-15-LO	Shooting Star	56-3675
Cessna	T-37B-CE	Tweet	56-3480
Northrop	T-38A-15-NO	Talon	59-1598
Lockheed	U-2D-LO		56-6721
De Havilland	U-6A-DH	Beaver	58-2039
North American	X-15A-2		56-6671

Kennedy administration's efforts to pull back from full-scale nuclear conflicts and to fight communism with conventional weapons. Also, much of the Air Force budget was now being spent on intercontinental guided missiles, stowed away in silos buried deep in the ground; these being considered the ultimate deterrent to all-out war. With the aid of hindsight more than thirty years later, it seems to have worked out for the best.

The growing size of the USAF missile effort could be studied at the numerous displays within the large hangar at Edwards, but for the aviation enthusiast, the only thing worth seeing were the aircraft displayed on the ramp. This continues true to the end of the century; long may it continue! The aircraft seen at this year's display are listed on the previous page.

Douglas TA-4F, 152848. A two-seat trainer version of the A-4E had been desired for several years, but Navy budgetary constraints continually delayed it until 1964. The demand for more A-4s in Vietnam finally convinced DoD to allow the last two A-4Es to be converted into TA-4Es. This required the fuselage to be lengthened by 28 inches; a considerable safety improvement was the addition of the new Douglas Escapac 1C-3 ejection seats. Lift spoilers were also added to the wing top trailing edge, giving much improved control in cross-wind landings (the long landing gear of the A-4 made it difficult to control in cross-winds). The first two aircraft flew as TA-4Es, but all of the improvements caused a change to the TA-4F designation. Production versions were then just becoming operational with VA-125, at NAS Lemoore, in California. A total of 241 TA-4E/Fs were finally built, this being the third production TA-4F.

Douglas A-4C, 145111, VA-195. Similar to the 1965 open house, a number of operational unit aircraft, especially from the Navy, were displayed this year. In addition to showing rarely seen unit markings at Edwards, later versions of aircraft tested there earlier were also seen. Here is a very clean A-4C from VA-195, showing the refueling probe on the right side of the nose, which had been lengthened by 9 in. This version of the A-4 became operational in February 1960, with its role now in the use of conventional, rather than nuclear, weapons.

(Left, Bottom) Douglas A-3B, 147650, VAH-2. The A-3B from VAH-2 was the 19th from last to be built, and differed from the earlier version in having the refueling probe, a new bomb director system behind a flat panel nose radome, and defensive electronic counter-measures equipment in a new dovetail fairing. These changes were gradually retrofitted to most of the earlier production A3D-2/A-3Bs.

Grumman A-6A, 149480, NATC. This was the first time we had seen the new Intruder at Edwards, this one visiting from the Naval Air Test Center. A very "chunky" looking airplane, it had been developed from Navy experience in the Korean War. Like the similar British Blackburn Buccaneer, the Intruder was developed to have high-subsonic performance at very low-levels, capable of penetrating below enemy radar defenses in all kinds of weather. The crew of two sat side-by-side (unlike the Buccaneer, which had tandem seating), while the two Pratt & Whitney J52 engines were mounted low on the fuselage sides; they had downward tilted exhaust pipes to shorten the take-off run. Up to 15,000 lb. of weapons could be carried externally on wing and fuselage pylons. The first A-6A flew in April 1960, and the type began operations in Vietnam in 1965. This is the 14th one built, and appears to be in use for testing weapon drops in view of the large number of low-drag bombs under the wings and fuselage.

(See also Top, Center p. 184) Grumman E-2A, 151707, VAW-11. Another new type seen for the first time at Edwards was this Grumman Hawkeye early-warning radar picket airplane. Designed as such from scratch, the type first flew in October 1960, designated as the W2F-1 (later changed to E-2A by the new DoD system). It was unusual in being a twin turboprop type, with a large diameter flat radome above the rear fuselage. No less than four vertical tails were required to meet both stability and control requirements caused by the airflow around the radome, plus the need to meet carrier storage height requirements. VAW-11 was the first unit to deploy the E-2A, in 1964, aircraft being detached as required by carriers operating in the Pacific area.

McDonnell F-4B, 152992, VF-121. Present among the Navy aircraft was this completely stock F-4B from VF-121. Like most of the other navy service aircraft of the period, it had very toned-down unit markings (just black letters and numbers). However, this would not last long under the exigencies of the war in Vietnam, where instant recognition in close combat soon outweighed the need for concealment by camouflage (exactly as happened in World War II).

Lockheed P-3A, 152617, PE. Yet another new type this year was the Lockheed P-3A Orion, this being a derivative of the Electra civil airliner. (The author's first design job in the USA in 1956 at Northrop had been design of portions of the rear empennage of the Electra, under subcontract to Lockheed. This was necessary to keep a group of 52 British engineers busy until their necessary USAF clearances were approved, this taking about three months). The prototype YP3V-1 flew in November 1959, and deliveries of aircraft to the Navy began in August 1962. Fitted with four Allison T56 turboprops, the Orion has given excellent service over the last thirty-five years, and will remain in service much longer. Note the high-power searchlight under the right wing, the rounded tips to the propellers (versus the square ones on the airliners), and the long MAD (Magnetic Anomaly Detector) anti-submarine tail-boom. Like the other Navy aircraft, the Orion is in a light gull gray and white color scheme, but as opposed to most of them, the white is on top of the fuselage, the rest of the aircraft being gray all over.

Chance Vought F-8E, 150672, VF-191, NM. This standard Navy Crusader is showing signs of the resurgence of colorful markings, with the red and white markings applied to the vertical tail and rudder. The type was giving excellent service in Vietnam at the time, pilots calling themselves "Gunfighters," due to the nose-mounted gun.

Douglas C-54M, 44-9093. Amazing, but the C-54 was still in use at Edwards, carrying a white top and tail. The only other piece of color was the USA flag on the tail. Note the "O" for a type more than ten years old (this C-54 was a mere 22 years old!).

Douglas C-118A-DO, 53-3265. Ostensibly the replacement for the C-54, this C-118 had merely supplemented it at Edwards. It was painted rather more plainly than the C-54, and had its flag at the top of the rudder. However, it still had the "O" in front of its serial number, but in this case it was only thirteen years old!

Vertol JCH-21B-VL, 53-4389. Once again, this old well-established type was on display, in its usual immaculate natural metal finish, with fresh fluorescent red panels applied.

Boeing NB-52B-BO-10, 52-008. This year, the static NB-52 was the second one, displayed with the rebuilt X-15A-2, 56-6671. Although the X-15 had been returned to Edwards with a 29 inch fuselage extension in February 1964, it did not fly until June 1964. Additional flight tests ran on through 1965, and the aircraft did not fly with its new external fuel tanks (empty on that flight) until November 1965. The X-15A-2 had been extensively modified since November 9, 1962, when it crashed during an emergency landing after engine problems. The intent of the rebuild was to push its performance out to the Mach 8 region, hence the additional fuel in the external tanks. In any event, the aircraft was to reach a speed of Mach 6.33 in November 1966, a world record at the time. These photos show the ever-increasing number of flight markings on the right side of the NB-52A, and the modifications to the rear end of the X-15A-2, including the helium tank mounted behind the vertical tail. The helium was required to pressurize the two external fuel tanks. Note how the extended fuselage caused the tail to move aft in relation to the cut-out in the B-52's wing. The X-15 was one of the most successful test programs ever carried out at Edwards, and the author counts himself as being very fortunate in having some connection with the test results and their use on other projects (when he was a project engineer on Douglas' hypersonic research programs).

Bell OH-13E-BF, 51-13793. No less than 490 of this small Bell helicopter were built, fitted with skid landing gear and side stretcher carriers. It had room for three persons in the cockpit, which was fitted with dual controls. It was powered by a 200 hp Franklin piston engine, giving a top speed of about 100 mph. This type was the sixth version of the Bell H-13 Sioux helicopter to be ordered, the first having being delivered in 1947, and paved the way for Bell to eventually become the largest supplier to the U.S. armed forces. It was among the smallest aircraft to be displayed at Edwards, but its usefulness can be gathered from it still being displayed with its much larger successors. One thing was for sure, it did not need much Army gloss olive drab paint to cover its structure!

(Center) Hiller OH-23D-UH, 55-3062. Powered like the Bell OH-13, by a piston engine—but this was a larger 250 hp Lycoming flat-six—the Raven was primarily used for utility and training duties. The first of the H-23Ds were delivered to the Army in 1957, and by the end of 1959, 130 of them had replaced earlier types at the Army Primary Helicopter School. Looking somewhat more substantial than the Bell H-13, it was also finished in gloss olive drab, but had large yellow markings, with a dash of fluorescent orange thrown in, to boot! The more solid appearance resulted in a top speed of only 95 mph, despite the extra 50 hp.

Lockheed T-33A-15-LO, 56-3675. As usual, one of the T-33s allocated to the Test Pilot School was displayed, this year being in natural metal finish all over. Note the small pod under the belly, normally used as a baggage container by pilots making overnight flights.

Cessna T-37B-CE, 56-3480. Also from the Test Pilot School, the T-37 was in highly polished natural metal, but had large areas of fresh fluorescent red over the fuselage, wing, and tail surfaces.

(Right, Bottom) Lockheed U-2D-LO, 56-6721. There were no apparent differences this year with the Edwards based U-2D, now in its third year of display.

Lockheed F-104A-10-LO, 56-0755. Again, no differences apparent on this long-time display aircraft.

North American XB-70A-NA, 62-001. The huge Mach 3 Valkyrie was present this year, both in the static display and in the flying display, the second prototype 62-207 having flown in July 1965. The changes between the first and second aircraft were not too obvious, the most visible being the lower black radome on the second one. There was a change on the wing to a 5 degree dihedral, but this was not easy to see. The second aircraft opened the flying display by flying overhead at over 30,000 feet, leaving a big sonic boom and a massive single vapor trail. The author visited the first aircraft in its hangar at Edwards a few weeks later, and walked all over the top of the wing around the vertical tail area. This gave a tremendous impression of the size of the aircraft, not to mention its height above the ground (this visit was in connection with the possibility of fitting a Mach 6 hypersonic research to the top of the XB-70. Powered by scramjets which would not run much below Mach 3, it was necessary to find a launch vehicle capable of flying at that speed. The XB-70 was the only available aircraft large and fast enough to do the job. It was not until 1977 that we found that Lockheed had tried such a launch system, the GTD-21 reconnaissance drone, on the A-11/YF-12 aircraft.They had run into launch problems, losing the launch aircraft, and eventually had to cancel the system).

(See also pp. 192-193) Lockheed YF-12A-LO, 60-6935. At last, the public were allowed to see close-up, the redoubtable Lockheed triple-sonic design. This example is the second of three YF-12A fighters built as variations of the original A-11. It had originated in November 1959, a month after OSD had canceled the North American stainless steel Mach 3 XF-108 fighter. In October 1964, the Air Force set requirements for the YF-12 that were higher than those which had been specified for the XF-108, including a combat radius of up to 1,200 nautical miles, a speed of Mach 3.0 plus, and greater maneuverability at high altitudes. As a result, two F-12 prototypes set nine speed and altitude records in May 1965 that were unbroken for many years. The YF-12s were tested at Edwards AFB by the 4786th Test Squadron until the unit was shut down in May 1972, a very long test period. A close look at the displayed aircraft revealed a host of very interesting features. The wing had an incredible thickness-to-chord ratio of only two and one half percent. The long slender fuselage and the two engine nacelles both featured forward chines, like the X-15 (the fuselage chines stopped well short of the nose, unlike the later SR-71s), while the fuselage was blended into the small delta wing. The twin vertical tails were very unusual at this time for a USA fighter design, together with three ventral fins (also lacking on the SR-71), while the sharp conical camber of the wing tips gave good airflow control. The main landing gear had no less than three wheels side-by-side on each leg, with very heat-resistant tires. The huge conical intake spikes proved to be one of the most difficult items to get right on the testing of the aircraft, as they were designed to keep the intake shock-wave precisely positioned at all speeds. They are angled down and inboard, to intercept the shock-wave from the aircraft's nose. Some aircraft were lost due to engines un-starting when the shock-wave was improperly positioned, resulting in tremendous asymmetrical thrust at very high speeds. A very high-speed electronic control system developed by Honeywell eventually cured this problem. The aircraft was built almost entirely of titanium, a very difficult material to work. A result of the structural design was that on the ground, the aircraft leaked fuel all over the place! A rather amusing sideline to the "Cold War" situation was that the best titanium sponge (basic raw material) came from the Soviet Union, and deliveries continued throughout this period. In fact, the F-4 and its J79 engines relied heavily on this Soviet material to obtain their performance throughout the Vietnam War! Incidentally, we now know that the world-startling MiG-25 fighter was designed to combat the A-11 and its various derivatives, the Soviet Air force being aware of their existence almost from the beginning.

FORCE

U.S. AIR FORCE

De Havilland U-6A-DH, 58-2039. The Beaver was by now a fixture in the U.S. Army, and this is a standard version, in the usual gloss olive drab overall, but still displaying lots of fluorescent red (one of the very few aircraft at this year's display with any such paint).

(Left, Center) Lockheed C-130B-05-LM, 57-525. The C-130A had eventually been upgraded by being fitted with more powerful Allison T-56-A-7 engines, of 4,050 eshp each. This allowed the operating weight to grow, and increased internal fuel capacity by some 1,820 gals. Designated the C-130B, the first of this version had flown in November 1958 at Marietta, Georgia (where all C-130 production had been transferred). The prototype is seen at Edwards, completely devoid of any unit markings, finished in the recent light gray color. Note the new four-bladed propellers, fitted to all C-130Bs and subsequent models, prior to the C-130J version.

Bell UH-1D-BF, 60-6029. The new version of the Iroquois had a much larger cabin capable of seating up to 12 troops in addition to the pilot, and a more powerful Lycoming T53 turboshaft engine, driving a larger diameter rotor. Some 2,008 were built, chiefly for use as troop carriers in the escalating Vietnam war. Noticeable is the very plain finish of gloss olive drab overall, with only yellow warning signs for the tail rotor. It does, however, still retain the national insignia; later in Vietnam service, the finish was changed to overall flat olive drab, with no national insignia.

(See also Top p. 196) Lockheed HC-130H-105-LM, 64-14852. Much more recent than the C-130B, was the HC-130H. The C-130H was an even more powerful version of the C-130E, fitted with 4,508 eshp Allison T56-A-15s, together with a redesigned wing center section to give greater service life (this new wing box was also retrofitted to many earlier versions of the C-130 family). The HC-130H version was built to replace the aging HC-54s used by the Air Rescue Service, and were fitted originally with the Fulton personnel recovery system yoke, as seen on the prototype aircraft. The red plastic un-inflated balloon was dropped to the person on the ground, attached to him, inflated, then the HC-130H would pick it up with the yoke on its nose. At the request of NASA, the HC-130Hs also had a Cook aerial tracking radar fitted to the top of the forward fuselage, to locate space capsules during reentry. This is the first HC-130H, of the 43 built.

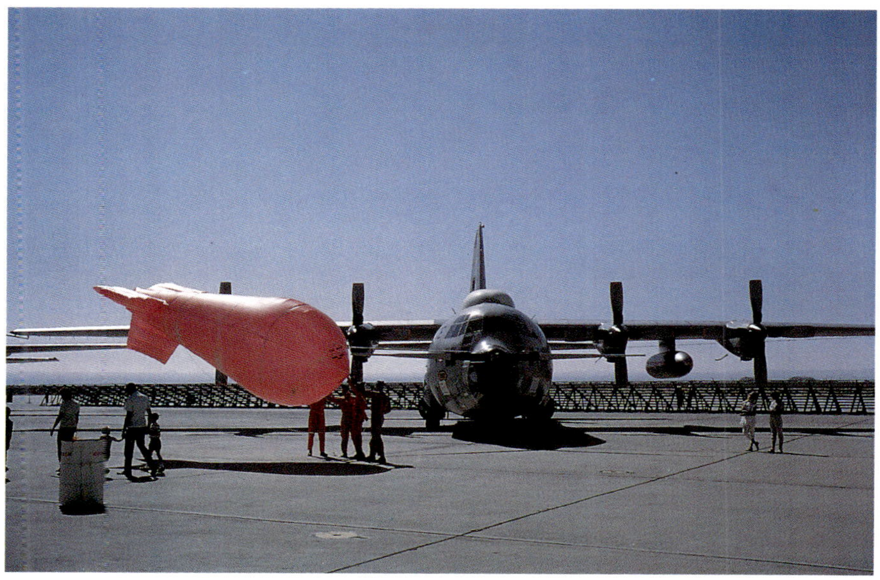

Boeing-Vertol JCH-47A-BV, 60-3449. First time seen at Edwards, was the Chinook Army transport, a larger outgrowth of the H-21 twin-rotor helicopter. The first one flew in September 1961, and deliveries to the Army began in August 1962. This is the second CH-47A, powered by two Lycoming T55 turboshaft engines producing 2,200 shp. Still painted in its flight test white finish with fluorescent red nose and tail areas, the helicopter stood out against the mostly drab finished Air Force and Navy aircraft.

Northrop T-38A-15-NO, 59-1598. This is the ninth production Talon, serving at Edwards in routine test pilot training, and chase operations at Edwards. It is seen in the now standard gloss white overall finish for the type. The only supersonic trainer designed as such, it has proved so successful that it is planned to continue its use into the 2020s.

Sikorsky JCH-3C-SI, 62-12577. In contrast to the large white Chinook, the new Sikorsky helicopter had originated from six Navy HSS-2s borrowed in 1962. Successful use of these led the Air Force to order the type as the CH-3C in November 1962, incorporating a new rear fuselage with a loading ramp for vehicles. The test helicopter displayed here is the first CH-3C, which had first flown in June 1963, 75 being procured. Their successors became known as the "Jolly Green Giant," after being used successfully in combat rescue missions in Vietnam. This prototype is in natural-metal finish overall

McDonnell F-4C-15-MC, 62-12199. The Phantom II was by now entering service with the USAF in large numbers, but the displayed aircraft was the first Air Force F-4C. Finished in the initial Navy flat light gull gray and gloss white finish, it shows off the multitude of small stenciled instructions all over the airframe.

Republic F-105F-1-RE, 62-4445. A real eye-opener, the F-105F was very fascinating this year, as it was the first combat aircraft to be displayed at Edwards in the new Vietnam camouflage color scheme. This consisted of a brown and two dark greens in an irregular pattern on all upper surfaces, with a very light gray on the lower surfaces. All colors were flat: introduced in 1965, this was the first time since 1943 that the Air Force had gone back to camouflage for its aircraft. Also noteworthy was the use of a very small national insignia, together with all other markings on the aircraft being in flat black. However, it had been found necessary to paint the aircraft serial number in larger white numbers on the vertical tail, thus continuing the everlasting battle between the need for aircraft concealment versus the need for instant recognition of its actual user (this was the case in all of the wars since WWI, and was to eventually result in the re-introduction of large white, visible, unit code letters later in the Vietnam war). Losses of the F-105D were by now beginning to mount in Vietnam, as it was the main USAF type used to carry the war to North Vietnam (although under very restrictive operational rules dictated from Washington, DC). The F-105F, as a two-seat aircraft, was required to teach already-proficient F-105D pilots how to use the very complex new weapon systems fitted to the D-model. The two-seater was to gain fame in other uses as the ultimate "Wild Weasel" F-105G SAM and radar killer later in the Vietnam war. This was the first time at Edwards that the USAF had allowed the public to view the cockpit of this very large fighter, providing a high viewing stand for this purpose (the author was able to sit in the cockpit of a F-105D at a later date, and noticed its great height above the ground).

4445

Northrop HL-10, NASA 804. Another new research vehicle shown this year was the strange-looking HL-10 lifting body vehicle (it was displayed in the hangar, so the author has chosen to show views of it taken by him at the roll-out at Northrop's Hawthorne plant in January 1966). This was the second of two such research vehicles built by Northrop, the first being the M2-F2 lifting body. The HL-10 had a conical forward fuselage blending into a flat bottom aft fuselage, with two outward canted vertical tails, plus a third center vertical tail: all three tails had rudders. Powered by rocket motors, the lifting bodies were carried by the venerable NB-52 launch vehicle, in a similar manner to the original Bell X-1 series. The tests were very successful and provided a great deal of data for the later Space Shuttle. The design was very ingenious in using existing systems to keep cost and development time down: for example, the nose gears used a modified T-39 nose gear, and the main gear used T-38 wheels on F-5 main gear struts. The rocket motor fitted after the first glide tests was a XLR-11, the same one used in the X-1 aircraft. After some modifications to improve aft-body airflow, the HL-10 finally reached a speed of Mach 1.86 at 67,310 feet, the fastest lifting body flight made: it also flew to an altitude of 90,303 feet. These achievements made the HL-10 the fastest and highest flying of the three lifting bodies (M2-F2, HL-10, and X-24).

Boeing NB-52A-BO, 52-0003. The by now aging NB-52A once again took part in the flying display, showing off more of the inscriptions on its right fuselage.

General Dynamics F-111A-CF, 63-9773 and 63-9774. These two static display F-111As were the eighth and ninth aircraft of the initial 23 development aircraft, and both were shown in the new USAF light gull gray and white finish (like the F-4Cs), complete with a unit badge on the first of the two. One was shown with the wings fully swept back, while the other had them swept fully forward. The very far aft "arm-pit" placement of the engine intakes, and the lack of any substantial boundary-layer bleeds were very noticeable to the practiced eye. Problems with the engine air intakes were to plague the F-111 program for several years, resulting in an industry-wide team effort to eventually resolve the problems. Other major problems arose from the use of a very complex, high strength, wing center section box, needed to carry the folding wings. The original USAF requirement for some 1,469 aircraft was eventually reduced to only some 520. Moreover, due to very different equipment fits, the USAF wound-up with F-111A, C, D, E, and F versions, each equipping a group, making all of them non-interchangeable. However, with tremendous expenditures of time and money, the F-111 finally proved its value in the Gulf War, rather than in the 25 year earlier Vietnam War (moral: never let politicians design any aircraft!).

LTV/Hiller/Ryan XC-142A, 62-5921. This new, very interesting aircraft was a further step along the path from the 1960 Hiller X-18 to the present-day Bell/Boeing V-22 Osprey, soon to enter service with the USAF and the Marines. Although given the service designation of XC-142 as a prototype cargo aircraft, it was too complex for its era. Some indication of its complexity could be gathered from its design and manufacture by no less than three separate companies. Ordered in 1962, it had made its first flight in September 1964, and did a great deal of development flying. Full of interesting features, many of which can be seen in these pictures, the XC-142 was the first tilt-wing aircraft to turn that technology into a useable form. Five were built, 62-5921/5925; the fourth one (62-5924) was eventually turned over to NASA, and given their identification number "522." Note the extent of travel on the horizontal stabilizer, which was also supplemented by a horizontal tail rotor, and that the wing position had been changed during the period of the display, showing its full travel. An aircraft of this type requires a very complex control system to keep it fully under control, especially during the transition from vertical to horizontal flight, and these problems require a powerful electronic system. This was one of the main reasons for the protracted development eventually leading to the Bell/Boeing V-22 Osprey.

XC-142
DANGER-PROPELLER

TRI-SERVICE

Lockheed YF-12A-LO, 60-6936. The second YF-12A also flew by, but in a very restrained fashion, obviously way below its true capabilities. There were no obvious differences between the two aircraft, apart from the nose markings showing three missile launches on the second aircraft.

North American XB-70A-NA, 62-207. As mentioned earlier, the second XB-70 opened the flying display with a very high vapor-trailing pass, creating a healthy sonic-boom in the process. Returning to normal fly-past altitudes, it flew past with its usual chase B-58 aircraft off its left wing, then made another run with aT-38A off its right wing. Its overall white finish showed up extremely well against the vivid blue sky, making a tremendous impression of size and speed. When it taxied past it was possible to see the 5 degree dihedral change from the first, static display, aircraft. It was with great regret that we heard on June 8, 1966, that the second aircraft had crashed near Barstow, California, during a flight photo session showing all of the latest GE-powered aircraft at Edwards. No one is sure of the cause of the crash, but it happened when a F-104 near the left wing tip rolled inverted over the top of the XB-70, knocking off most of that aircraft's twin verticals. All pilots involved in the flight were very experienced test pilots, and unfortunately two were killed in the crash. Beside the loss of such an expensive and valuable aircraft, the crash resulted in major career losses for many people involved at Edwards AFB. Among their many achievements, the two XB-70s and the A-11/YF-12/SR-71 family have given the USAF more flight time at Mach 3 than any other military service in the world.

U.S. AIR FORCE

Convair B-58A-CO, 55-0662. The long time-resident B-58A used as the chase aircraft for the XB-70 test flights is seen as it taxies in after its fly-past with the second XB-70.

Conclusion

We have finished our look at ten years of aircraft displayed at Edwards AFB, from 1957 to 1966, and seen what tremendous strides were made in those years in aircraft design and performance. Unfortunately, a great deal of the pressure for these advancements in military aircraft came from the very long Cold War, with its outbursts into "Hot Wars," such as in Korea and Vietnam. However, much of this accelerated development also paved the way for major advances in commercial aviation, especially the development of comfortable, fast, and affordable jet transports. One other major change over the years has been the ever-decreasing number of major aircraft manufacturers: indeed, many of those whose aircraft have been described above have passed from the scene, including such companies as Chance Vought, Grumman, Republic, Fairchild, Hiller, North American, Vertol, Convair, General Dynamics, Martin, Douglas, and McDonnell. We hope that you have enjoyed this review of the industry's products tested at Edwards AFB: the author hopes to do another volume, covering the period between 1967 and 1991, the latter being the year when he retired from the industry.

Lockheed C-141A-LM, 61-2779, MATS. The C-141A was fast becoming the key aircraft in MATS cargo fleet, and here one of the test aircraft at Edwards is seen taxiing past, in its highly polished natural metal finish.

Appendix 1
Northrop Model 176 Mach 2+ All-Weather Interceptor

As mentioned in the caption to the 1960 photographs of the Northrop YT-38A, it had been rumored that Northrop had tendered to the XF-108 interceptor program, and had been awarded the T-38 program as a consolation for losing that effort. Whatever the truth, Northrop had been working on developments of the F-89 interceptor, plus many advanced new designs of much greater performance. Here is an artist's impression of the massive Northrop N-176 Mach 2+ interceptor, together with a color cutaway showing its main internal systems.

Chief thing to notice is the striking resemblance of the aircraft configuration to the earlier N-156 configuration, using a straight wing and horizontal tail almost identical in taper. However, the extraordinary thing about this configuration is the use of no less than eight GE J85 turbojets. The manner in which these were mounted as a block of four units exhausting into a common afterburner was a unique installation. This would have led to some major maintenance problems in servicing the inner engines, plus the possible damage from one major engine failure. The author does not know the exact configuration of the proposed Northrop Mach 3 interceptor, but it may have been derailed by similar system configuration problems. However, later development of the F-5 and F-20 fighters showed that there was plenty of potential performance in the N-176 basic aerodynamic configuration.

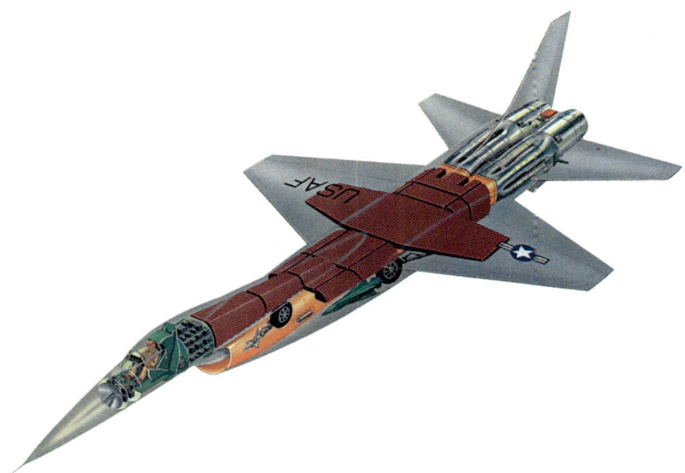

Appendix 2
McDonnell F4H-1
Spin Test Recovery System Mechanism

After the Northrop engineering group had finished designing the major components of the McDonnell F4H-1, as discussed in the 1959 captions to the prototype aircraft photographs, the author was tasked with designing a mechanism to control the deployment of spin recovery chutes for the upcoming spin tests of the F4H-1. The plural use of chutes is correct, because the McDonnell test pilot who was going to make the spin test flights had had some bad experiences with single chutes on previous aircraft installations.

The desired use of two totally independent chutes led to some major problems, not least because of the large diameter chute needed to recover such a heavy fighter. The only possible location for the two chutes and their associated mechanisms was at the extreme rear of the fuselage. However, this location was right above the blast from the two big afterburning J79 engines, a very hot spot.

After much discussion, the author decided to use a redesigned version of the WW 2 glider tow release mechanism, because of its outstanding reliability. The biggest problem was in getting two such devices into a box attached to the rear longerons of the F4H-1. Stress checks of the design soon showed that there was insufficient room to install enough bolts to take the internal loads and hold the unit to the aircraft. The only solution was to weld the entire box as a unit and provide enough external attachments to bolt it to the airframe. This led to a major problem in that the only material strong enough to carry the loads at the temperatures in the area could not be welded, according to all available data!

Fortunately, the author found a very capable manufacturing engineer at Northrop who felt that he could solve the problem, despite almost everyone else saying that it couldn't be done. Fortunately, we persevered in our efforts and he solved the problem by using a combination of a new welding rod material, together with doing the welding in a sealed chamber filled with an inert gas (the welder had to wear suitable breathing equipment).

Next problem was how to read the completed welds to ensure their strength. This was solved by sending the unit after every weld up to the UCLA medical X-ray unit, as they had the only unit powerful enough to read such large welds. It all worked well, due to the dedicated efforts of all concerned.

When the completed mechanism was tested in a special rig in the shops at Northrop, it raised a lot of attention. This was because, when the expected loads from a deployed parachute were released by the mechanism, it sounded just like a 75mm canon being fired at close proximity! The completed unit was shipped to St. Louis and the author heard nothing more about it. However, the F4H-1 completed its spin tests satisfactorily, so he had to assume that it worked as required. The photograph shows the complete unit being assembled into the special rear end unit holding the mechanisms and dual parachutes; persons around the unit were the shop foreman, George Whorley (ex-Supermarine engineer) who had designed the structure, a shop mechanic, and the author (all wearing shop safety glasses).

Appendix 3
Northrop YT-38A Spin Recovery System

When the author was transferred from the completed McDonnell F4H-1 project, he went to the Northrop T-38A design group. After doing very interesting work on the main and nose landing gears, he worked on other projects until asked to design the spin recovery system for the YT-38A prototype.

Wind tunnel tests had indicated that the T-38 configuration could get into very fast rotating spins, with a high rate of sink, chiefly due to the mass of the aircraft being concentrated into the long fuselage. By contrast, the small span wings contained little mass, this contributing to a very high roll rate. The majority of military aircraft designed up to this time had used long span wings (straight or swept), with fuel placed in them, together with somewhat shorter fuselage lengths. Data gained from the rather imprecise tunnel tests tended to show that the forward "G" forces on the pilot could be high enough to kill him (by stopping him from breathing)!

With this rather alarming data bank, tests were run on the centrifuge at USC to check the forces that the Northrop test pilot, Jack Wells, could take in the forward direction (from behind). He did very well, but the predicted forces were even higher. The author was tasked with designing the complete spin recovery system, together with testing it during high speed taxi tests at Edwards AFB. A similar mechanism to the F4H-1 was developed, but only for one chute, Jack Wells apparently having enough confidence in that being sufficient. A large diameter parachute was mounted on top of the aft fuselage, above the release mechanism. In view of the very high drag of the chute, it was decided that the release mechanism would be attached to the beefed-up airframe by explosive bolts. This would allow the pilot to get rid of the deployed chute if all else failed.

Initial tests on the Edwards runway proved to be somewhat disastrous after a few test runs, when the pilot chute deployed inadvertently as the aircraft was being held on the brakes, the engines being run up to full power. This pulled out the nylon main chute, which then proceeded to melt in the jet blast! Investigation showed that a "safety" cover had been installed without the author's knowledge. This cover was installed over the switch used by the pilot to fire the cartridge which deployed the pilot chute and got it out into clean air about 100 feet away from the aircraft.

When further tests showed that the system was ready for use, the T-38 proceeded with the spin flight tests. More than 200 flights were then flown without the pilot being able to get it into a spin, and we thought that we had finally designed an unspinnable aircraft. However, Jack finally got the T-38 into a very violent spin. To do this he had to, in an inverted flight at a climb angle of about 45 degrees, push the

stick hard forward into the corner of the cockpit (a most unnatural pilot command)! The T-38 toppled and then proceeded to spin inverted, pitching up and down at a high rate while rolling very rapidly from side-to-side until it rolled right side up. It then continued to spin flat with a high rotation rate and a very high sink rate. At this point, unable to recover the aircraft with any aircraft control inputs, Jack deployed the spin chute. This stopped the spin and the aircraft ended up sinking vertically under the chute, both engines having flamed out in the preceding wild gyrations. Almost all of the people present during this test flight thought that we had lost the aircraft. Jack then released the chute with its normal control system, re-lit the engines, and proceeded to make a safe landing.

Later test flights showed that there were many other ways of getting into a spin. The aircraft was capable of recovering from most of these, but the spin chute was required for several more recoveries. Moral: don't spin any modern aircraft without being prepared to eject!

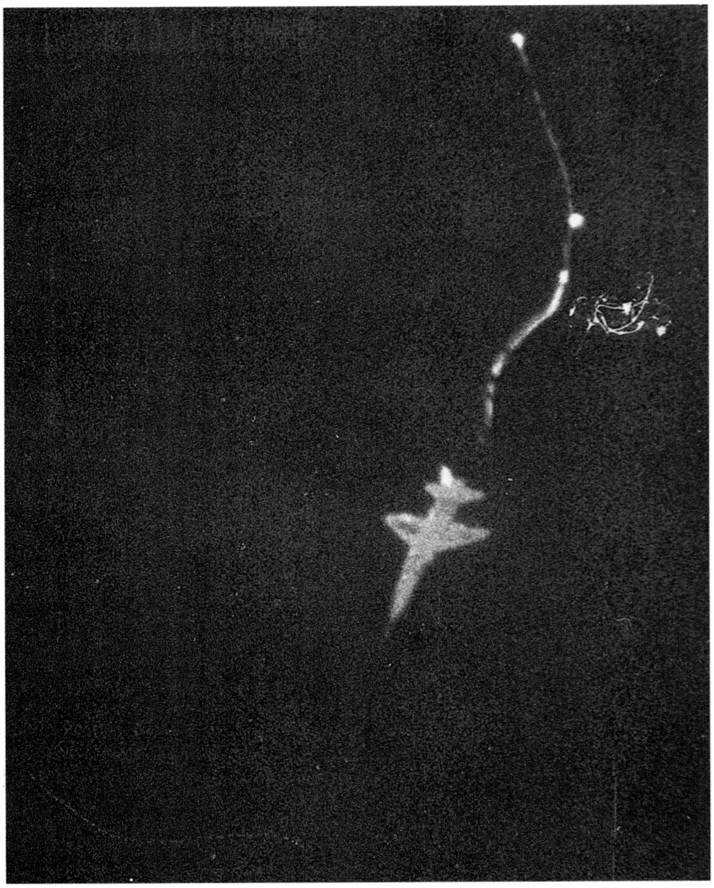

Appendix 4
Northrop N-156F/F-5 Drag Chute System

Northrop's persistence in designing light-weight fighters finally paid-off when the company was awarded a contract for three prototype fighter versions of the N-156T/T-38 design. In fact, the fighter version was conceived in engineering AD (advanced design) before the trainer N-156T version, but the trainer version was ordered first by the Air Force. The new fighter aircraft were designated as the N-156F by Northrop. The author was assigned to design the drag parachute (normally shortened to chute in the industry) system, starting with the initial preliminary design passed down from AD. This used a long small-diameter cylindrical container, with a long tapered door, to house the drag chute above the aft boat-tail fuselage.

Unfortunately, no amount of effort by the author and Irvin Parachute Company personnel could get the required 24 foot diameter ribbon chute into this container (the chutes were designed to strict requirements established at Wright Field). Eventually, in desperation, the author talked to the key aerodynamic engineer for the airplane and asked if it would be possible to put a short, wide box-like container with a flat plate door above the end of the fuselage. The aerodynamicist replied that no-one really knew exactly what the boundary layer was doing at the rear end of the aircraft at high trans-sonic speeds, so he figured that we could use the suggested box container, and it would not be possible to accurately define any possible base-drag difference.

Armed with this information, the author went ahead with a new design which allowed a very simple installation of the drag chute and its operating controls. Unfortunately, the original AD designer took great exception to this change to his design and demanded a review of the whole system by the program manager. That manager listened very carefully to the pro and con arguments and finally decided to implement the box design. That's how the F-5 family ended up with its unique drag chute installation.

We later found out that the other competing fighters with long circular container (such as the French Mirage III) used an entirely different type of chute, consisting essentially of two rectangular sheets of fabric at right angles to one-another. That chute provided ample drag, but it was very unstable when deployed and was not then approved by Wright Field for aircraft use.

The chute was required to give the N-156F the short landing run needed for operation from short, dirt-type fields. The photographs show the chute being used on the Edwards AFB runway, and on the grass alongside the runway at the old Hughes Aircraft Culver Field. The latter use was during a competition for a short-field attack/fighter aircraft which included a version of the Douglas A4D aircraft, fitted with a special drag chute system.

At that time, all drag chute systems had to be tested at Wright Patterson AFB, near Dayton, Ohio, using a special

version of the old Douglas A-26B Invader. This aircraft, an A-26B-45-DL, serial number 44-34137, had its outer wings cut off just outboard of the engine nacelles, and was fitted with extra large wheels and brakes (to stop it if the planned chute deployment didn't work), together with a special drag chute housing at the end of the fuselage. The photograph shows the N-156F chute deployed successfully on its first run at a speed of 115 knots.

Although drag chutes were used on all of the USAF fighters of the period, their use has diminished on current aircraft, because they added complications in operational use. They required special chute packing facilities, including drying ones for use in wet weather, trained personnel to recover the chute after use, pack them, etc.. Larger brakes and wheels, thrust reversers, and large horizontal tail surfaces with a lot of movement provide the shorter landing runs when required on modern aircraft.

Appendix 5
Sud Caravelle VII with GE Engines Flight at Edwards AFB, March 1961

In 1960, Sud Aviation of France had sold more than 60 Caravelles and were very anxious to sell the type in the United States. They entered into an agreement with Douglas for them to act as the sales agent for the Caravelle in major portions of the world. As a result a Caravelle III was sold to General Electric (GE) to be fitted with aft-fan CJ-805s fitted in place of the Rolls-Royce Avons. It was delivered to GE on July 18, 1960 and converted to the type VII five months later. The author flew in a press flight of the Caravelle III on August 31, 1960, before it was converted and was able to sample it after it had been converted, during another press flight, this time at Edwards AFB on March 29, 1961.

The considerably increased thrust of the CJ-805s, 16,100 lb. versus the 11,400 lb. of the Avons, gave a big improvement to the short-field capability of the Caravelle. It was intended to lengthen the fuselage by 39 inches and increase the weight by 10,000 lb. in the production Caravelle VII. However, the major American airlines showed little or no interest in the type (except for United Airlines buying 20 of the Caravelle VI version) and the prototype was converted back to a Caravelle III in 1963 and sold to Air France. Major result of the whole exercise was that Douglas decided that there was a market for a short-haul jet in the United States and started development of what finally became the DC-9. It was also another step in the eventual development of the French Airbus company many years later.

GE's press flight of the converted aircraft was carried out on one of those nasty windy days common at Edwards in this time of the year, leading to considerable turbulence at around 10,000 feet altitude. We took off and climbed at a very steep

climb angle to 30,000 feet at the north-west side of Edwards. After we had reached that altitude, the aircraft was leveled off and flown to the limiting Mach number of 0.77. Stalls were then demonstrated after power was retarded to idle, occurring at about 112 knots and 20 degrees angle of attack in the clean condition. Stalls in the landing configuration resulted in speeds about 5 knots lower.

The author was then invited to take the left seat by GE pilot "Swede" Davis. Seat adjustment was simple and all controls fell well to hand. Forward view was somewhat restricted by the small windows of the de Havilland Comet I nose used on the early Caravelles (larger windows were fitted to all later production Comet 4Bs and Caravelles). The controls felt very solid, with the aileron and elevator forces feeling much lighter

than the decidedly heavy rudder. A sudden drop in right engine power to idle without warning by Davis, caused only a very slight wing-drop and a decrease in speed. The change was so small that it was barely perceptible in the control system, and very easy to correct.

Despite the apparent heaviness of the controls, the aircraft was easy to handle, allowing fully co-ordinated turns to be made with high angles of bank. A simulated landing and go-around showed that the Caravelle VII accelerated very quickly up to its maximum flaps down speed of 190 knots, requiring a sharp pull-up to reduce speed while the aircraft was cleaned up. Unfortunately, the weather was beginning to deteriorate, causing cancellation of the planned landing demonstrations. The Caravelle rode the increased turbulence well on the return trip to Edwards, landing smoothly despite the severe cross-wind conditions prevailing.

All-in-all, the re-powered Caravelle came out as very attractive airplane, but it was unable to gain any orders, as mentioned earlier, and disappeared into history (the author was saddened to hear of "Swede" Davis' death in May 1965, when doing spin tests on the obscure German Hamburger HFB 320 swept-forward wing corporate jet in Spain, a reminder of the dangers test pilots face in their very demanding profession).

(The author's much longer original article on this flight was published by Flight magazine, London, England, in their issue dated 8 June 1961).

Appendix 6
Northrop T-38 Flight at Palmdale, January 1961

The Northrop T-38A Talon has proved to be the most successful supersonic trainer to ever enter service with any air force. Making its first flight in April 1959, the T-38 had not required any major changes to its design and went into large-scale production, some 1,189 being built. As a result, almost all pilots wanted to get their hands on it, to sample true supersonic flight.

The USAF finally gave permission for the author to fly in the T-38, but not until he had obtained a high-altitude chamber "green card", allowing flight at over 18,000 feet altitude. This was duly obtained in the chamber at Edwards AFB, and the author was informed that he could make his flight in a T-38 at Palmdale, where the aircraft were finally assembled prior to first flight. However, it was made a condition for this permission that in no way was it to be implied in the resulting article that the USAF favored the T-38 and its N-156F fighter version over any other type being offered for service to NATO countries (this was the time when Lockheed, backed by the USAF, were trying to sell the F-104 upgrade throughout Europe, and the USAF did not want to help sales of the N-156F which might take business away from the F-104. It became a very bitter struggle, with Lockheed winning all of the early major sales).

After agreeing to this condition, the author was told to report to Palmdale on January 28, 1961. He was slightly surprised to find that the aircraft allocated was 59-1603, because this was fitted with up-rated J85-5 engines, producing nearly 4,000 lbs. of afterburning thrust each, and was being tested for an attack on the world time to-altitude record. The Northrop test pilot was Hank Chouteau and he briefed me on the forthcoming flight profile. During the aircraft walk-around, the author noticed a small saddle fairing added between the two engine nozzles. Hank explained that this had been added to cure a small buffet, noticeable only at military power. This was the 16th T-38 produced and the first production T-38A-20, finished in the standard gloss white, with broad bands of fluorescent red on the wings, fuselage, and empennage.

The author was then strapped into the rear, instructor's, cockpit and thoroughly briefed on the use of the oxygen system and the ejection seat procedures. He found that there was ample room in the cockpit for his cameras and equipment, and that the view forward was superb. After starting the engines and receiving permission to taxi to the active runway, Hank then told the author to steer the aircraft out, using

the nosewheel steering; this proved to be easy but required somewhat high pedal forces. We closed the canopies, and checked that they were locked properly. Satisfied, Hank said that he was going to show how fast the engines could be accelerated on the ground. It only took 5 seconds for each engine to reach military power.

The take-off was extremely impressive for its acceleration, an aft pull on the stick at 135 knot lifted us off after a 2,300 feet roll. Gear and flaps were retracted immediately to not exceed their limiting speeds (240 kn for gear and 300 kn for flaps). No trim change was noticeable in either case. Hank held the T-38 down at about 60 feet and let it gather speed until we were doing 450 kn at the end of the 12,000 foot runway. The stick came back for a 2g pullup, and we climbed at about 50 degrees straight up to 45,000 feet. Mach number was over 0.9 during the climb and the initial rate of climb was greater than 30,000 feet/minute. We leveled off at 45,000 feet a scant four minutes after brakes-off, a most impressive performance!

Hank then demonstrated the roll rate at about 200 degrees/second, handed the aircraft over to me with instructions to fly down to 36,000 feet and turn to the east. The aircraft was extremely responsive to control inputs, with no signs of pilot-induced oscillations. Turns at Mach 0.8 were made wholly on the ailerons, and it was easy to fly the aircraft wherever it was placed on the attitude director. At 36,000 feet, Hank accelerated the aircraft from Mach 0.8 to 1.25, only sign of going supersonic being a slight flick of the Mach meter needle at 1.0. Cutting the afterburners felt like hitting a brick wall, with very sudden deceleration.

Hank made a practice ILS approach to Palmdale to check some of the aircraft's instrumentation, as this was only its fourth flight, then we came in for the landing. Flaps down caused a high rate of sink, simulating modern fighters. Gear was lowered at 200 kn, and at 175 kn buffeting began, becoming most noticeable at 155 kn over the fence. Touchdown was at 135 kn with the nose held up, the wide track gear giving a very positive touchdown. Smooth, even, braking resulted in a landing roll of about 3,000 feet.

In summary, the T-38 showed outstanding acceleration and rate of climb, exceptional handling qualities throughout the speed range, superb visibility and roomy cockpits. All-in-all, a remarkable airplane whose qualities will keep it flying well into the next century, more than 60 years after its first flight.

91603
TF-603

91603
TF-603

Appendix 7
North American T-39A Flight at Edwards AFB

Early in May 1961, the USAF invited the author to take an orientation flight at Edwards AFB in the new North American T-39A Sabreliner. The pre-flight briefing given by Col. Scott Anderson covered the background history of the aircraft and its basic characteristics.

The T-39 was designed to fit an USAF requirement for an off-the-shelf, FAA civil certificated light twinjet transport, basically to replace its aging fleet of piston engine Beech C-45s. Under this UTX (Utility Transport Experimental) program, North American designed and built a prototype aircraft which made its first flight in September 1958. The flight tests went very smoothly, and the aircraft completed the USAF Phase II evaluation in December 1958. USAF ordered seven T-39As in January 1959, plus two T-39B radar & navigation trainers in October 1959. A further 35 T-39s were ordered in 1960.

First flight of the production aircraft was made in June 1960, and the required FAA certification program began in November 1960.

Col. Anderson then introduced USAF Captain Davy Benefield, pilot in charge of the flight, and we proceeded out to the main ramp where the fifth production T-39A, serial number 59-2872, was waiting for us. Basically, the T-39 was very similar in size and weight to the F-86 Sabre fighter, but with a lower maximum Mach number of .765. Walk-around showed a close-to-the-ground configuration, with the swept wing fitted with full span leading edge aerodynamically operated slats. A large F-100-like speed brake was fitted under the forward fuselage aft of the twin wheel nose gear. The two aft mounted Pratt & Whitney podded J60 engines of 3,000 pounds thrust each had their intakes above and over the wing trailing edge.

A trimmable horizontal stabilizer with elevators sat near the top of the aft fuselage. All control surfaces were fitted with electric powered trim tabs. The cockpit had very large bird proof forward windows giving superb visibility to the two pilots. The aircraft could be flown from the left seat only, all controls being fully accessible. The main cabin was fitted with two aft-facing and two forward-facing seats; small pear-shaped windows gave rather limited vision to the passengers.

On the flight day, it was typical Edwards spring weather, with a strong north-west cross- wind blowing across the main runway. After startup, the aircraft proved to be easy to taxi and responsive to the slightly heavy nose-wheel steering. Takeoff roll was about 2,500 feet, lifting off at 117 knots and then climbing out to 35,000 feet, taking 11.45 minutes to reach 31,000 feet. Cruise was at 260 knots IAS (Mach .76) at 35 000 feet, with a fuel flow of 800 pounds per hour at 90% rpm. This gave us an available 3.5 hr flight time, after our first 30 minutes.

The speed brake proved to be very effective, with little trim change, but it proved to be very noisy. When the author flew the aircraft, it could not be stalled above 15,000 feet due to the high angle of attack at higher altitudes causing engine compressor stalls and flame-outs. (duly rectified on later aircraft). Using the speed brake, we then descended to 15,000 feet and proceeded to stall the aircraft in both clean and landing configurations. The stall characteristics were very good, without any tendency to drop a wing; particularly noteworthy was how the slats on both wings moved in and out perfectly evenly. North American's long experience with similar slats on the F-86 and F-100 swept wings really showed up here. Stall in the clean configuration was at about 95 knots and with flaps and gear down, at about 90 knots. After recovery the engines only took about 4 to 5 seconds to accelerate from idle to military power.

Electing to make a straight-in approach to Edward's 15,000 ft runway required quite a bit of rudder and aileron to counteract the very nasty northwest cross-wind. Differential throttle was not used due to the close proximity of the thrust lines. Approach was made at about 160 knots, using the speed brake as required. Flaps were extended at 140 knots, reducing airspeed to about 130 knots. The considerable nose down trim caused by flap extension was taken out by using the stabilizer electric trim manually. Approach was made at about 115 knots, with the nose held up. The strong cross-wind required a crab-like approach until just before touch down, then the aircraft was lined up with the runway, using rudder and aileron. This produced a satisfyingly smooth touch down at about 100 knots. Nose wheel steering easily kept the aircraft straight on the runway.

Although North American originally had no intention of producing a civil version of the T-39, demand eventually resulted in the commercial Sabreliner and both military and civil versions of the aircraft went on to long successful careers. All-in-all, the T-39A proved to be a very pleasant aircraft to fly, with great visibility from the cockpit and excellent handling characteristics.

Appendix 8
North American XB-70

North American XB-70 roll-out at Palmdale, May 1964, and first flight from Palmdale to Edwards AFB, Sep. 1964. Edwards aircraft at Palmdale airshows in 1964 and 1965.

The long awaited roll-out of the XB-70 finally took place on May 11, 1964, in front of a very large invited group of visitors. First appearance of the aircraft was its very high off the ground cobra-like nose peering out of its hanger, then slowly the rest of the huge Mach 3 aircraft appeared in its gleaming white finish. There were many cheers and gasps from the 5,000-strong crowd when the entire aircraft appeared out of the hanger. Unfortunately, the program had been canceled except for the two prototypes, partly due to the huge costs involved and also due to the earlier shooting down of the U-2 over the Soviet Union. This showed that the Soviets had missiles that could attain the altitude that the XB-70 was designed to fly at (however, it is of interest that they were never able to shoot down the other USA Mach 3 jet, the SR-71, despite trying for a very long time).

The photographs show various aspects of the XB-70 design, then are followed by those of its first flight, from Palmdale to Edwards some months later, on September 21, 1964. The take off was uneventful, as the huge aircraft lifted off, escorted by a Vertol HH-21 helicopter (for ground rescue if necessary), a Convair B-58, and a flock of Northrop T-38s. It had been planned that the press photographers would be transported from Palmdale to Edwards by bus, about a one hour drive. This would give ample time for all to get into place for the landing. However, the best laid plans of men and mice do go astray at times, and this unfortunately was one of those times.

As the buses were driving along the road west of the base, about three miles from the press vantage point, someone exclaimed that "there it comes!" We later found that a series of problems in trying to retract the landing gear led the two pilots to a decision to cut the planned flight short. The buses stopped at the request of the press members and we all piled out to watch the approach and landing. Apparently, the author was the only one to even try to photograph the approach, and the results are presented here for the first time (other than the USAF and North American photographers in place on the base). The XB-70 made a good, smooth approach, but when it touched down, the brakes on the left main

gear locked up (or were before landing), and the resulting friction burnt off and melted the left main gear tires and wheels (the smoke can be seen in the last of those photographs).

Despite this rocky start, the XB-70 went on to make a total of 83 flights before being put on display at the USAF Museum, Wright-Patterson AFB, Dayton in February 1969.

* * *

We decided to add a few significant photographs taken at two airshows at Palmdale in November 1964 and October 1965 to finish this work.

We rarely saw any of the US Army aircraft fly at the Edwards displays, so it was good to get a close view of this Grumman OV-1A Mohawk, taxiing close-by before takeoff.

This was the first time that we saw a camouflaged USAF tactical aircraft, resulting from the build-up of the war in Vietnam. This was North American F-100D-90-NA, 56-325?, painted in the two greens and tan on the upper surfaces and light gray on the lower surfaces. Note the tiny national insignia and the hard to read serial numbers on the vertical tail. It made a tremendous contrast with the Thunderbirds natural-metal F-100s.

Again, we did not see the LTV/Hiller/Ryan XC-142A fly at Edwards, so it was very interesting to see how well it flew at Palmdale.

In those days, the USAF Thunderbirds did not fly at the Edwards shows, so it was great to see the team with their distinctive markings fly at Palmdale. In worsening weather, we see the team taking off. Note that the leader has just lit his afterburner, with a great display of flame. Seen are F-100D-20-NAs 55-3507, 55-3582, 55-3506, 55-3561.

Appendix 9
New Military Aircraft Designation System, Department of Defense, 1963

As related in some of the captions for 1963, Secretary of Defense McNamara, ordered a new system to be adopted immediately, for designating all military aircraft. A person who was very proud of his capability to retain huge amounts of information, he lost his temper during a session with Congress, when he temporarily could not remember that the McDonnell F4H-1 and F-110A were the same aircraft. Thus, he ordered that all aircraft of the same type should have a common designation for use in any of the military services. Unfortunately, the Air Force and the Navy had evolved their own systems for designating their aircraft during the previous forty years or so (the Marine Corps used the Navy system for their aircraft, which were usually ordered by the Navy).

In the event, new designations were assigned to currently operational aircraft of the Navy, Marines, Coast Guard and Army, while most of the Air Force aircraft retained their previous designations. However, the situation became very confusing even for Air force aircraft, because they were ordered to assign a totally new set of, low number, designations to all new aircraft. This despite the fact that the new designations would often duplicate those given to previous Air Force aircraft (though most of them had been used in the 1920s and 1930s). As can be imagined, this has led to all kinds of confusing duplication of aircraft designations, causing many problems for service logistic personnel, not to mention aviation historians. Perhaps the worst consequence was that the new system destroyed any logical sequence of designations, leading to such anomalies as the Air Force C-5 transport coming into service after the C-141, and the F-4 coming into service before the F-111.

Designation Standards

Designation standards used in the new system are shown in the example below. The YA3D-2P, a service test version of the Douglas Skywarrior, is now designated the YRA-3B-DL in accordance with the following breakdown:

1. *Status:* Prototype Testing
2. *Modified Mission:* Reconnaissance
3. *Basic Mission:* Attack
4. *Design:* Third
5. *Series:* Second or B
6. *Source:* Douglas Long Beach

Y R A - 3 B - DL

Following is an explanation of the designation standards and a list of symbol definitions used in the new designation system:

Status

G—Permanently grounded
J— Special test, temporary
N—Special test, permanent
X—Experimental
Y— Prototype
Z—Planning

2. Modified Mission

A—Attack
C—Cargo/Transport
D—Drone director
E—Special electronic installation
H—Search or rescue
K—Tanker
L—Cold weather
M—Missile carrier
Q—Drone
R—Reconnaissance
S—Anti-submarine search
T—Trainer
U—Utility
V—Staff
W—Weather

3. *Basic Mission*

A—Attack
B—Bomber
C—Cargo/Transport
E—Special electronic
F—Fighter
H—Helicopter
O—Observation
P—Patrol
S—Anti-submarine Search
T—Trainer
U—Utility
V—VTOL or STOL
X—Research
Z—Airship

4. *Design Number*

The design number is a digit issued consecutively to identify each major model.

5. *Series Symbol*

Each variation of the basic major model is identified by a series symbol, i.e., A for first variation, B for second variation, etc.

6. *Source*

The source, or manufacturer's code, is a two-letter symbol used to identify the prime or subcontractor manufacturer.

Following is a list of major companies with their assigned symbol and address:

Symbol Manufacturer and Address

AE Aeronca Aircraft Corp., Middletown, Ohio
BH Beech Aircraft Corp., Wichita, Kansas
BF Bell Helicopter Corp., Ft. Worth, Texas
BC Bell Aerosystems Co., Buffalo, New York
BN Boeing Co., Renton, Washington
BO Boeing Co., Seattle, Washington

BV Boeing Co. (Vertol Division), Morton, Pennsylvania
BW Boeing Co., Wichita, Kansas
CE Cessna Aircraft Co., Wichita, Kansas
CF Convair, Ft. Worth, Texas
CO Convair, San Diego, California
DH de Havilland Aircraft of Canada, Toronto, Ontario
DM Doman Helicopter, Inc., Danbury, Connecticut
DL Douglas Aircraft Co., Inc., Long Beach, California
DO Douglas Aircraft Co., Inc., Santa Monica, California
DT Douglas Aircraft Co., Inc., Tulsa, Oklahoma
FA Fairchild Aircraft Division, Hagerstown, Maryland
GO Goodyear Aircraft Co., Akron, Ohio
GI Grand Central Aircraft Engine Co., Tucson, Arizona
GR Grumman Aircraft Engine Corp., Bethpage, L.I., New York
GY Gyrodyne Co. of America, Inc., St. James, L.I., New York
HE Helio Aircraft Corp., Norwood, Massachusetts
HI Hiller Helicopter Corp., Palo Alto, California
HU Hughes Tool Co., San Diego, California
KA Kaman Helicopter Corp., Windsor Locks, Connecticut
LM Lockheed Aircraft Corp., Marietta, Georgia
LO Lockheed Aircraft Corp., Burbank, California
MA Martin Co., Baltimore, Maryland
MD Martin Co., Denver, Colorado
MF Martin Co., Orlando, Florida
MC McDonnell Aircraft Corp., St. Louis, Missouri
ND Noorduyn Aviation Co., Ltd., Montreal, Quebec
NA North American Aviation, Inc., Inglewood, California
NH North American Aviation, Inc., Columbus, Ohio
NI North American Aviation, Inc., Downey, California
NO Northrop Aircraft, Inc., Hawthorne, California
PI Piasecki Aircraft Corp., Philadelphia, Pennsylvania
PA Piper Aircraft Corp., Lockhaven, Pennsylvania
RE Republic Aviation Corp., Farmingdale, L.I., New York
RY Ryan Aeronautical Co., San Diego, California
SW Schweitzer Aircraft Corp., Elmira, New York
SI Sikorsky Aircraft Division, Stratford, Connecticut
TA Taylorcraft Aviation Corp., Alliance, Ohio
VO Chance Vought Aircraft, Dallas, Texas

Department of Defense Aircraft Designations

Note: The source symbols listed below indicate the prime contractor. In many cases, the same specific aircraft was also built at other locations or by other manufacturers.

ATTACK SERIES

Current Designation	Former Designation	Service	Current Designation	Former Designation	Service
A-1E-DL	AD-5	Navy	RA-3B-DL	A3D-2P	Navy
EA-1E-D	AD-5W	Navy	TA-3B-DL	A3D-2T	Navy
EA-1F-DL	AD-5Q	Navy	YA-4A-DL	Y/A4D-1	Navy
A-1G-DL	AD-5N	Navy	A-4A-DL	A4D-1	Navy
A-1H-DL	AD	Navy	YA-4B-DL	Y /A4D-2	Navy
A-1 J-DL	AD-7	Navy	A-4B-DL	A4D-2	Navy
A-2A-NH	AJ-1	Navy	YA-4C-DL	Y/A4D-2N	Navy
YA-3A-DL	Y/A3D-1	Navy	A-4C-DL	A4D-2N	Navy
A3A-DL	A3D-1	Navy	A4E-DL	A4D-5	Navy
YEA-3A-DL	Y/A3D-1Q	Navy	YA-5A-NH	Y/A3J-1	Navy
EA-3A-DL	A3D-1Q	Navy	A-5A-NH	A3J-1	Navy
YRA-3A-DL	Y /A3D-1P	Navy	A-5B-NH	A3J-2	Navy
A-3B-DL	A3D-2	Navy	A-5C-NH	A3J-3	Navy
EA-3B-DL	A3D-2Q	Navy	A-6A-GR	A2F-1	Navy
YRA-3B-DL	Y/A3D-2P	Navy	EA-6A-GR	A2F-1H	Navy

NOTE: Future New Aircraft will be designated starting at Design No. -7A.

BOMBER SERIES

Current Designation	Former Designation	Service	Designation	Current Designation	Former Service
VB-26B-DL	Same	AF	B-52D-BO	Same	AF
B-26B-DL	Same	AF	B-52E-BO	Same	AF
TB-26B-DL	Same	AF	B-52F-BO	Same	AF
B-26C-DL	Same	AF	B-52G-BO	Same	AF
B-26C-DL	GB-26C	AF	B-52H-BO	Same	AF
RB-26C-DL	Same	AF	B-57A-MA	Same	AF
TB-26C-DL	Same	AF	RB-57A-MA	Same	AF
UB-26J-DL	JD-1	Navy	B-57B-MA	Same	AF
DB-26J-DL	JD-1D	Navy	RB-57B-MA	Same	AF
YB-26K-DL	New	AF	B-57C-MA	Same	AF
TB-29A-BC	Same	AF	TB-57C-MA	Same	AF
B-45C-NA	Same	AF	RB-57D-MA	Same	AF
RB-45C-NA	Same	AF	RB-57D-MA	Same	AF
B-47B-BW	Same	AF	B-57E-MA	Same	AF
DB-47B-BW	Same	AF	TB-57E-MA	Same	AF
TB-47B-BW	Same	AF	B-58A-CF	Same	AF
WB-47B-BW	Same	AF	YRB-58A-CF	Same	AF

B-47E-BW	Same	AF
EB-47E-BW	Same	AF
RB-47H-BW	ERB47H	AF
RB-47K-BW	Same	AF
TB-50D-BO	Same	AF
WB-50D-BO	Same	AF
RB-50F-BO	Same	AF
KB-50J-BO	Same	AF
KB-50K-BO	Same	AF
B-52A-BO	Same	AF
B-52B-BO	Same	AF
B-52C-BO	Same	AF
TB-58A-CF	Same	AF
RB-66A-DL	Same	AF
B-66B-DL	Same	AF
RB-66B-DL	Same	AF
RB-66C-DL	Same	AF
B-66D-DL	Same	AF
WB-66D-DL	Same	AF
B-66E-DL	Same	AF
XB70A-NA	XB-70	AF

NOTE:- New Bomber Aircraft will be designated starting at Design No. -IA.

CARGO/TRANSPORT SERIES

Current Designation	Former Designation	Service
C-45G-BH	Same	AF
C-45H-BH	Same	AF
RC-45H-BH	Same	AF
TC-45H-BH	Same	AF
RC-45J-BH	SNB-5P	Navy
TC-45J-BH	SNB-5	Navy
C-46A-CU	Same	AF
C-46D-CU	Same	AF
TC-46D-CU	Same	AF
C-47A-DL	Same	AF
C-47H-DL	R4D-5	Navy
EC-47H-DL	R4D-5Q	Navy
HC-47A-DL	SC-47A	AF
LC-47H-DL	R4D-5L	Navy
RC-47A-DL	Same	AF
SC-47H-DL	R4D-5S	Navy
TC-47H-DL	R4D-5R	Navy
C-54A-DL	Same	AF
VC-54N-DL	R5D-1Z	Navy
C-54P-DL	R5D-2	Navy
VC-54P-DL	R5D-2Z	Navy
C-54D-DL	Same	AF
C-54Q-DL	R5D-3	Navy
EC-54D-DL	AC-54C	AF
HC-54D-DL	SC-45D	AF
TC-54D-DL	Same	AF
VC-54D-DL	Same	AF
VC-54Q-DL	R5D-3Z	Navy
RC-54V-DL	R5D-3	Navy(C.G.)
C-54E-DL	Same	AF
C-54R-DL	R5D-4R	Navy
EC-54U-DL	R5D-4	Navy(C.G.)
C-54G-DL	Same	AF
C-54S-DL	R5D-5	Navy
VC-54G-DL	Same	AF
VC-54S-DL	R5D-5Z	Navy
C-54M-DL	Same	AF
VC-118B-DL	R6D-1Z	Navy
C-119J-FA	MC-119	AF
C-119C-FA	Same	AF
C-119F-FA	R4Q-2	Navy
C-119G-FA	Same	AF
C-119J-FA	Same	AF
C-121A-LO	Same	AF
C-121C-LO	Same	AF
C-121J-LO	R7V-1	Navy
TC-121C-LO	Same	AF
EC-121K-LO	WV-2	Navy
VC-47A-DL	Same	AF
VC-47H-DL	R4D-5Z	Navy
C-47J-DL	R4D-6	Navy
EC-47J-DL	R4D-6Q	Navy
LC-47J-DL	R4D-6L	Navy
SC-47J-DL	R4D-6S	Navy
TC-47J-DL	R4D-6R	Navy
VC-47J-DL	R4D-6Z	Navy
VC-47B-DL	Same	AF
C-47D-DL	Same	AF
EC-47D-DL	AC-47D	AF
HC-47D-DL	SC-47D	AF
RC-47D-DL	Same	AF
TC-47D-DL	Same	AF
VC-47D-DL	Same	AF
C-47E-DL	Same	AF
TC-47K-DL	R4D-7	Navy
C-54T-DL	R5D-5R	Navy
C-97A-BO	Same	AF
C-97C-BO	Same	AF
C-97D-BO	Same	AF
VC-97D-BO	Same	AF
KC-97F-BO	Same	AF
KC-97G-BO	Same	AF
C-97J-BO	Same	AF
C-117A-DL	Same	AF
VC-117A-DL	Same	AF
C-117B-DL	Same	AF
VC-117B-DL	Same	AF
C-117C-DL	Same	AF
C-117D-DL	R4D-8	Navy
LC-117D-DL	R4D-8L	Navy
TC-117D-DL	R4D-8T	Navy
VC-117D-DL	R4D-8Z	Navy
C-118A-DL	Same	AF
C-118B-DL	R6D-1	Navy
VC-118A-DL	Same	AF
EC-121H-LO	Same	AF
EC-121L-LO	WV-2E	Navy
EC-121M-LO	WV-2Q	Navy
C-123B-FA	Same	AF
C-123H-FA	Same	AF
C-124A-DL	Same	Navy
C-124C-DL	Same	AF
C-130A-LM	Same	AF
C-130A-LM	GC-130A	AF
RC-130A-LM	Same	AF
C-130B-LM	Same	AF

YEC-121K-LO	WV-2	Navy	C-130F-LM	GV-1U	Navy
RC-121D-LO	Same	AF	HC-130B-LM	SC-130B	AF
WC-121N-LO	WV-3	Navy	HC-130G-LM	SC-130B	Navy(C.G.)
RC-121E-LO	Same	AF	KC-130F-LM	GV-1	Navy
VC-121E-LO	Same	AF	LC-130F-LM	C-130BL	Navy
C-121G-LO	Same	AF	C-130D-LM	Same	AF
C-130E-LM	Same	AF	C-140A-LM	Same	AF
HC-130E-LM	SC-130E	AF	C-140B-LM	Same	AF
C-131A-CO	Same	AF	VC-140B-LM	Same	AF
VC-131A-CO	Same	AF	C-140C-LM	UV-1	Navy
C-131B-CO	Same	AF	C-141A-LM	Same	AF
C-131D-CO	Same	AF	XC-142A-VO	Same	AF
C-131F-CO	R4Y-1	Navy			
C-131E-CO	Same	AF			
C-131G-CO	R4Y-2	Navy			
C-133A-DL	Same	AF			
C-133B-DL	Same	AF			
RC-135A-BN	Same	AF			
C-135A-BN	Same	AF			
KC-135A-BN	Same	AF			
C-135B-BN	Same	AF			
VC-137A-BO	Same	AF			
VC-137C-BO	Same	AF			

Because of the size of the design numbers for Cargo Aircraft, this series will start over at -1A. Following are new assignments made under series of numbers:

C-1A-GR	TF-1	Navy
EC-1A-GR	TF-1Q	Navy
C-2A-GR	New	Navy
C-3		
TC-4B-GR	New	Navy

NOTE: Future New Cargo/Transport Aircraft will be designated starting at Design No. -5A.

SPECIAL ELECTRONIC INSTALLATION SERIES

Current Designation	Former Designation	Service
EA-1E-GR	E-2A-GR	Navy
WF-2	W2F-1	Navy

NOTE: Future New Special Electronic Installation Aircraft wil be designated starting at Design No. -3A.

FIGHTER SERIES

Current Designation	Former Designation	Service	Current Designation	Former Designation	Service
F-80C-LO	Same	AF	F-101A-MC	Same	AF
F-84F-RE	Same	AF	YRF-101A-MC	Same	AF
RF-84F-RE	Same	AF	RF-101A-MC	Same	AF
F-84G-RE	Same	AF	F-I01B-MC	Same	AF
F-86D-NA	Same	AF	TF-101B-MC	Same	AF
F-86F-NA	Same	AF	F-101C-MC	Same	AF
F-86H-NA	Same	AF	RF-101C-MC	Same	AF
F-86L-NA	Same	AF	F-101F-MC	Same	AF
F-89D-NO	Same	AF	TF-101F-MC	Same	AF
F-89H-NO	Same	AF	F-102A-CO	Same	AF
F-89J-NO	Same	AF	TF-102A-CO	Same	AF
F-100A-NA	Same	AF	YF-102C-CO	Same	AF
F-100C-NA	Same	AF	F-104A-LO	Same	AF
DF-100C-NA	Same	AF	CF-104A-LO	Same	AF
F-100D-NA	Same	AF	F-104B-LO	Same	AF
F-100F-NA	Same	AF	F-104C-LO	Same	AF
DF-100F-NA	Same	AF	F-104D-LO	Same	AF
F-104G-LO	Same	AF	MF-1C-NH	FJ-3M	Navy
RF-104G-LO	Same	AF	DF-1D-NH	FJ-3D2	Navy
TF-104G-LO	Same	AF	YF-1E-NH	Y/FJ-4	Navy
F-104J-LO	Same	AF	F-1E-NH	FJ-4	Navy
TF-104J-LO	Same	AF	YAF-1E-NH	YFJ-4B	Navy
F-105B-RE	Same	AF	AF-1E-NH	FJ-4B	Navy
F-105D-RE	Same	AF	F-2C-MC	F2H-3	Navy
F-106A-CO	Same	AF	F-2D-MC	F2H-4	Navy
F-106B-CO	Same	AF	YF-3B-MC	Y/F3H-2	Navy
YF-106C-CO	Same	AF	F-3C-MC	F3H-2	Navy
F-111A-CF	TFX	AF	MF-3B-MC	F3H-2M	Navy
F-111B-CF	TFX	Navy	F-3C-MC	F3H-2N	Navy

Because of the size of the design numbers for fighter aircraft, this series will start over at -1. Following are new assignments made under new series of numbers.

F-1C-NH	FJ-3	Navy
DF-1C-NH	FJ-3D	Navy
YF-6A-DL	Y/F4D-1	Navy
F-6A-DL	F4D-1	Navy
YF-7A-CO	Y/F2Y-1	Navy
YF-8A-VO	Y/F8U-1	Navy
F-8A-VO	F8U-1	Navy
DF-8A-VO	F8U-1D	Navy
QF-8A-VO	F8U-1KD	Navy
YRF-8A-VO	Y/F8U-1P	Navy
RF-8A-VO	F8U-1P	Navy
TF-8A-VO	F8U-1T	Navy
F-8B-VO	F8U-1E	Navy
YF-8C-VO	Y/F8U-2	Navy
F-8C-VO	F8U-2	Navy
YF-8D-VO	Y/F8U-2N	Navy
F-8D-VO	F8U-2N	Navy
YF-8E-VO	Y/F8U-2NE	Navy
F-8E-VO	F8U-2NE	Navy
DF-9E-GR	F9F-5KD	Navy
F-4A-MC	F4H-1F	Navy
F4B-MC	F4H-1	Navy
RF-4B-MC	F4H-1P	Navy
F-4C-MC	F-110A	AF
RF-4C-MC	RF-110A	AF
F-5A-NO	N-156	AF
F-5B-NO	N-156	AF
F-9F-GR	F9F-6	Navy
DF-9F-GR	F9F-6D	Navy
QF-9F-GR	F9F-6K	Navy
QF-9G-GR	F9F-6K2	Navy
F-9H-GR	F9F-7	Navy
F-9J-GR	F9F-8	Navy
YAF-9J-GR	Y/F9F-8B	Navy
AF-9J-GR	F9F-8B	Navy
YTF-9J-GR	Y/F9F-8T	Navy
TF9J-GR	F9F-8T	Navy
RF-9J-GR	F9F-8P	Navy
F-10A-DL	F3D-1	Navy
F-10B-DL	F3D-2	Navy
EF-10B-DL	F3D-2Q	Navy
MF-10B-DL	F3D-2M	Navy
TF-10B-DL	F3D-2T2	Navy
YF-11A-GR	Y/F11F-I	Navy
F-11A-GR	F11F-I	Navy

NOTE: Future New Fighter Aircraft will be designated starting at Design No. -12A.

HELICOPTER SERIES

Current Designation	*Former Designation*	*Service*
UH-1-BF	HU-1	Army
UH-1A-BF	HU-1A	Army
UH-1B-BF	HU-1B	Army
UH-1D-BF	HU-1D	Army
UH-1E-BF	HU-1E	Navy
UH-2A-KA	HU2K-1	Navy
UH-2B-KA	HU2K-1U	Navy
YSH-3A-SI	Y/HSS-2	Navy
SH-3A-SI	HSS-2	Navy
VH-3A-SI	HSS-2Z	Navy
OH-4A-BF	HO-4	Army
OH-5A-HI	HO-5	Army
OH-6A-HU	HO-6	Army
OH-13E-BF	H-13E	Army
OH-13G-BF	H-13G	Army
OH-13H-BF	H-13H	Army
CH-21A-BV	H-21A	AF
CH-21B-BV	H-21B	AF
CH-21B-BV	H-21B	Army
OH-23B-HI	H-23B	Army
OH-23C-HI	H-23C	Army
OH-23D-HI	H-23D	Army
OH-23F-HI	H-23F	Army
UH-25B-BV	HUP-2	Navy
UH-25C-BV	HUP-3	Navy
XH-33A-BF	Same	AF
CH-34A-SI	H-34A	Army
CH-34C-SI	H-34C	Army
LH-34D-SI	HUS-1L	Navy
YSH-34G-SI	Y/HSS-1	Navy
SH-34G-SI	HSS-1	Navy
UH-13H-BF	H-13H	AF
UH-13J-BF	H-13J	AF
OH-13K-BF	H-13K	Army
TH-13L-BF	HTL-4	Navy
TH-13M-BF	HTL-6	Navy
TH-13N-BF	HTL-7	Navy
UH-13P-BF	HUL-1	Navy
HH-13Q-BF	HUL-1G	Navy (C.G.)
UH-13R-BF	HUL-1M	Navy
UH-19B-SI	H-19B	AF
HH-19B-SI	SH-19B	AF
UH-19C-SI	H-19C	Army
UH-19D-SI	H-19D	Army
CH-19E-SI	HRS-3	Navy
UH-19F-SI	HO4S-3	Navy
HH-19G-SI	HO4S-3G	Navy (C.G.)
UH-34E-SI	HUS-1A	Navy
HH-34F-SI	HUS-1G	Navy (C.G.)
SH-34H-SI	HSS-1F	Navy
YSH-34J-SI	Y/HSS-1N	Navy
SH-34J-SI	HSS-1N	Navy
CH-37A-SI	H-37A	Army
CH-37B-SI	H-37B	Army
CH-37C-SI	HR2S-1	Navy
XH-40A-BF	Same	AF
NH-41A-CE	YH-41	Army
UH-41A-CE	H-41	AF
HH-43A-KA	H-43A	AF
HH-43B-KA	H-43B	AF
UH-43C-KA	HUK-1	Navy
OH-43D-KA	HOK-1	Navy

UH-34D-SI	HUS-1	Navy
VH-34D-SI	HUS-1Z	Navy
CH-46C-BV	HC-1A	Army
CH-47A-BV	HC-1B	Army
XH-48A	HX/H-1	AF
QH-50A-GY	DSN-1	Navy
QH-50B-GY	DSN-2	Navy
QH-50C-GY	DSN-3	Navy
XH-51A-LO	New	Navy
HH-62-SI	HU2S-1G	Navy (C.G.)
CH-53A-SI	New	Navy
CH-46A-BV	HRB-1	Navy
XCH-46B-BV	HX/H-2	AF

NOTE: Future New Helicopters will be designated starting at Design No.-54A.

OBSERVATION SERIES

Current Designation	Former Designation	Service
O-1A-CE	L-19A	Army
O-1B-CE	OE-1	Navy
O-1C-CE	OE-2	Navy
TO-1D-CE	TL-19D	Army
O-1E-CE	L-19E	Army

NOTE: Future New Observation Aircraft will be designated starting at Design No. -2A

PATROL SERIES

Current Designation	*Former Designation*	*Service*
P-2D-LO	P2V-4	Navy
P-2E-LO	P2V-5F	Navy
DP-2E-LO	P2V-5FD	Navy
EP-2E-LO	P2V-5FE	Navy
SP-2E-LO	P2V-5FS	Navy
P-2F-LO	P2V-6	Navy
MP-2F-LO	P2V-6M	Navy
TP-2F-LO	P2V-6T	Navy
P-2G-LO	P2V-6F	Navy
YP-2H-LO	Y/P2V-7	Navy
P-2H-LO	P3V-7	Navy
SP-2H-LO	P2V-7S	Navy
LP-2J-LO	P2V-7LP	Navy
YP-3A-LO	Y/P3V-1	Navy
P-3A-LO	P3V-1	Navy
QP-4B-CO	P4Y-2K	Navy
P-5A-MA	P5M-1	Navy
SP-5A-MA	P5M-1S	Navy
TP-5A-MA	P5M-1T	Navy
P-5B-MA	P5M-2	Navy
SP-5B-MA	P5M-2S	Navy

NOTE: Next Design No. will be -7A.

ANTI-SUBMARINE SERIES

Current Designation	Former Designation	Service
S-2A-GR	S2F-1	Navy
YS-2A-GR	Y/S2F-1	Navy
TS-2A-GR	S2F-1T	Navy
S-2B-GR	S2F-1S	Navy
S-2C-GR	S2F-2	Navy
RS-2C-GR	S2F-2P	Navy
S-2D-GR	S2F-3	Navy
S-2E-GR	S2F-3S	Navy

Note: next Design No. will be -3A.

TRAINER SERIES

Current Designation	Former Designation	Service
T-1A-LO	T2V-1	Navy
YT-2A-NH	Y/T2J-1	Navy
T-2A-NH	T2J-1	Navy
T-2B-NH	T2J-2	Navy
T-28A-NA	Same	Navy
T-28A-NA	Same	AF
T-28B-NF	Same	Navy
ET-29C-CO	AT-29C	AF
T-33A-LO	Same	AF
DT-33A-LO	game	AF
RT-33A-LO	Same	AF
WT-33A-LO	Same	AF
T-33B-LO	TV-2	Navy
DT-33B-LO	TV-2D	Navy

T-28B-NF	Same	AF	DT-33C-LO	TV-2KD	Navy
DT-28B-NF	T-28BD	Navy	T-34A-BH	Same	AF
T-28C-NF	Same	Navy	T-34B-BH	T-34B	Navy
T-28D-NF	Same	AF	YT-34B-BH	Y/T-34B	Navy
YAT-28E	New	AF	T-37A-CE	Same	AF
T-29A-CO	Same	AF	T-37B-CE	Same	AF
VT-29A-CO	Same	AF	T-38A-NO	Same	AF
T-29B-CO	Same	AF	YT-38A-NO	Same	AF
VT-29B-CO	Same	AF	T-39A-NA	Same	AF
VT-29C-CO	Same	AF	T-39B-NA	Same	AF
T-29C-CO	Same	AF	T-39D-NA	T3J-1	Navy
T-29D-CO	Same	AF	T-40A-LO	Same	AF
VT-29D-CO	Same	AF			

NOTE: Next New Trainer Design will start at No.-41A.

UTILITY SERIES

Current Designation	Former Designation	Service
U-1A-DH	Same	Army
U-1B-DH	UC-1	Navy
U-2A-LO	Same	AF
U-3A-CE	Same	AF
U-3B-CE	Same	AF
U-4A-AD	Same	AF
U-4B-AD	Same	AF
U-6A-DH	L-20A	AF
U-6A-DH	L-20A	Navy
U-6A-DH	L-20A	Army
U-7A-PA	L-21	AF
U-8D-BH	L-23D	Army
RU-8D-BH	RL-23D	Army
U-8E-BH	L-23E	Army
U-8F-BH	L-23F	Army
U-9B-AD	L-26B	Army
U-9C-AD	L-26C	Army
RU-9D-AD	RL-26D	Army
U-1OA-HE	L-28A	AF
U-11A-PA	UO-1	Navy
HU-16A-GR	SA-16A	AF
HU-16B-GR	SA-16B	AF
HU-16C-GR	UF-1	Navy
LU-16C-GR	UF-1L	Navy
TU-16C-GR	UF-1T	Navy
HU-16D-GR	UF-2	Navy
HU-16E-GR	UF-2G	Navy (C.G.)

NOTE: Utility series will continue with design No.-17A.

VTOL & STOL SERIES

Current Designation	Former Designation	Service
OV-1A-GR	AO-1A	Army
OV-1B-GR	AO-1B	Army
OV-1C-GR	AO-1C	Army
CV-2A-DH	AC-1	Army
CV-2B-DH	AC-1A	Army
XV-3A-RY	None	Army
XV-4A-LM	VZ-10	Army
XV-5A-RY	VZ-11	Army
XV-6A-HA	VZ-12	Army
CV-7A-DH	AC-2	Army

NOTE- Next V Design No. will start at -8A.

RESEARCH SERIES

Current Designation	Former Designation	Service
X-13A-RY	X-13	AF
X-14A-BC	X-14	AF
X-15A-NA	X-15	AF
X-19A-CU	X-19	AF
X-20A-BO	X-20	AF
X-21A-NO	X-21	AF
X-22A-BC	New	AF

NOTE: Next Design No. is -23A.

AIRSHIP SERIES

Current Designation	Former Designation	Service
EZ-1B-GO ZPG-2W	Navy	
SZ-1B-GO ZPG-2	Navy	
EZ-1C-GO ZPG-3W	Navy	

NOTE: Next Design No. is -2A.

Fighter Aircraft Flyaway Cost per Production Aircraft
(in then-year dollars)

Type	Total Cost	Airframe	Engines (installed)	Electronics	Ordnance	Armament
Lockheed F-80	$95,000					
FP-80A	107,796	75,957	21,584	4,195	2,335	3,715
F/TF-80C (T-33)	93,456	62,050	21,192	5,536	-	4,678
Republic F-84B	163,994	-	-	-	-	-
F-84C	147,699	-	-	-	-	-
F-84D	212,241	139,863	41,654	7,165	-	23,559
F-84E	212,241	139,863	41,654	7,165	-	23,559
F-84G	237,247	150,846	41,488	4,761	2,719	37,433
F-84F	769,330	562,715	146,027	9,623	9,252	41,713
RF-84F	667,608	482,821	95,320	21,576	4,529	63,632
North American F-86A	178,408	101,528	52,971	7,576	-	16,333
F-86E	219,457	145,326	39,990	6,358	4,138	23,645
F-86F	211,111	140,082	44,664	5,649	3,047	17,669
F-86H	582,493	316,360	214,612	6,831	17,117	27,573
F-86D	343,839	191,313	75,036	7,085	419	69,986
Northrop F-89B	1,085,882	950,298	90,364	4,870	-	40,350
F-89C	797,202	612,533	95,110	10,557	4,519	74,483
F-89D	801,602	598,439	101,954	11,392	1,857	87,960
F-89H	988,884	536,748	105,697	10,094	998	335,347
F-89J	988.884	536,748	105,697	10,094	998	335,347
Lockheed F-94A	258,123	193,721	45.227	4,014	-	15,161
F-94B	196,248	123,422	31 336	7,635	2,947	30,908
F-94C	534,073	380,755	90 147	7,058	518	55,595
North American F-100A	1,014,910	748,259	217,390	8,549	20,807	19,905
F-100C	633,181	439,323	173,554	12,050	21,125	12,125
F-100D	697,029	448,216	162,995	10,904	8,684	66,230
F-100F	804,444	577,023	143,527	13,677	3,885	66,332
McDonnell F-101A	2,906,373	2,364,143	429,016	25,249	15,300	72,665
F-101C	1,276,145	805,022	287,764	61,079	441	123,839
RF-101A	1,604,963	1,150,903	288,466	32,566	591	132,457
RF-101C	1,276,145	803,022	287,764	61,079	441	123,839
F-101B	1,754,066	1,105,034	332,376	52,770	1,001	262,885
Convair F-102A	1,200,000	744,258	210,308	9,208	525	219,876
TF-102A	1,500,000	1,135,018	144,474	11,365	1,192	173,777
Lockheed F-104A	1,700,000	1,026,859	624,727	3,419	29,517	19,706
F-104B	2,400,000	1,756,388	336,015	13,258	59,473	231,996
F-104C	1,500,000	862,235	473,729	5,219	44,684	91,535
F-104D	1,500,000	873,952	27[illegible],148	16,210	70,067	269,014
F-104G	1,420,000	1,251,000	169,000	in airframe	in airframe	in airframe
Republic F-105B	5,649,543	4,914,016	328,797	141,796	32,021	232,913
F-105D	2,140,000	1,472,145	244,412	19,346	19,346	167,621
F-105F	2,200,000	1,524,000	290,000	251,000	21,000	154,000
Convair F-106A	4,700,000	2,090,000	274,000	1,300,000	102,000	950,000
F-106B	4,900,000	2,200,000	274,000	1,350,000	24,000	1,089,000
General Dynamics F-111A		8,200,000	4,304,000	1,354,000	1,688,000	7,000 925,000
F-111E	9,200,000	4,756,000	1,511,000	1,945,000	7,000	1,060,000
FB-111A	9,800,000	4,201,000	1,735,000	2,550,000	-	1,342,000
F-111D	8,500,000	3,895,000	1,229,000	2,530,000	6,000	844,000
F-111F	10,300,000	5,097,000	2,026,000	1,711,000	6,000	1,529,000
McDonnell F-4C	1,900,000	1,388,725	317,647	52,287	-	139,706
RF-4C	2,300,000	1,679,000	276,000	293,000	-	73,000
F-4D	1,700,000	1,018,682	260,563	262,101	6,817	133,430
F-4E	2,400,000	1,662,000	393,000	299,000	8,000	111,000
Northrop F-5A	756,000	578,000	155,000	11,000	2,700	9,300
F-5B	1,200,000	856,000	218,000	22,000	6,000	31,000
RF-5A	890,000	676,000	175,000	33,000	2,700	3,300
F-5E	2,100,000	1,625,000	426,000	47,000	5,000	17,000
Lockheed F-12	Between 15,000,000 and 18,000,000 each			for 100 aircraft, if they had been ordered.		

All Data Extracted from the "Encyclopedia of U. S. Air Force Aircraft and Missile Systems, Vol. 1, Post-World War II Fighters, 1945-1973", by Marcelle Size Knaack, Office of Air Force History, 1978.